The Zero-Sum Society

THE ZERO-SUM SOCIETY

Distribution and the Possibilities for Economic Change

Lester C. Thurow

Basic Books, Inc., Publishers NEW YORK

Library of Congress Cataloging in Publication Data

Thurow, Lester C
The zero-sum society.

Includes bibliographical references and index.
1. United States–Economic conditions–1971–
2. United States–Economic policy–1971–
3. Income distribution–United States. I. Title.
HC106.7.T49 330.973'092 79–2758
ISBN: 0–465–09384–1

Printed in the United States of America
DESIGNED BY VINCENT TORRE
10 9 8 7 6 5 4

Contents

The Zero-Sum Society

Chapter 1

An Economy That No Longer Performs

AFTER DECADES of believing in their economic invulnerability, Americans were jolted by the 1973–74 Arab oil embargo. The actions of a few desert sheiks could make *them* line up at the gas pump and substantially reduce *their* standard of living. Sudden economic vulnerability is disconcerting, just as that first small heart attack is disconcerting. It reminds us that our economy can be eclipsed.

When the shutdown of a major oil exporter for just a few months in 1979 once again resulted in the convulsions of gas lines, it was possible to ask whether that first mild heart attack was not the harbinger of something worse. Seemingly unsolvable problems were emerging everywhere–inflation, unemployment, slow growth, environmental decay, irreconcilable group demands, and complex, cumbersome regulations. Were the problems unsolvable or were our leaders incompetent? Had Americans lost the work ethic? Had we stopped inventing new processes and products? Should we invest more and consume less? Do we need to junk our social welfare, health, safety, and environmental protection systems in order to compete? Why were others doing better?

Where the U.S. economy had once generated the world's highest standard of living, it was now well down the list and slipping farther each year. Leaving the rich Middle East sheikdoms aside, we stood fifth among the nations of the world in per capita GNP in 1978, having been surpassed by Switzerland, Denmark, West Germany, and Sweden.[1] Switzerland, which stood first, actually had a per capita GNP 45 percent larger than ours. And on the outside, the world's fastest economic runner, Japan, was advancing rapidly with a per capita GNP only 7 percent below ours. In our entire history we have never grown even half as rapidly as the Japanese.

While the slippage in our economic position was first noticed in the 1970s, our economic status was actually surpassed (after just half a century of delivering the world's highest standard of living) by Kuwait in the early 1950s.[2] Kuwait was ignored, however, as a simple case of a country inheriting wealth (oil in the ground) rather than earning it. We failed to remember that our supremacy had also been based on a rich inheritance of vast mineral, energy, and climatic resources. No one inherited more wealth than we. We are not the little poor boy who worked his way to the top, but the little rich boy who inherited a vast fortune. Perhaps we had now squandered that inheritance. Perhaps we could not survive without it.

Of course, one can always argue that things are not really as bad as they seem. Since many goods are not traded in international markets and may be cheaper here than abroad, per capita GNP may paint too pessimistic a picture of our relative position. A group of American economists argued in 1975 that we still had the highest real standard of living among industrialized countries.[3] What we lost in per capita GNP to the two or three countries that were then ahead of us, we more than made up in terms of lower living costs.

Whether this is still true today depends upon changes in the *terms of trade*—the amount of exports that you have to give up to get a given amount of imports. In Switzerland, for example, oil cost less in 1978 than it did in 1975.[4] While the dollar price of oil is up, the value of the Swiss franc is up even more. Thus fewer domestic goods have to be given up to buy a given quantity of oil. The country's GNP simply buys more than it did. In countries like Switzerland, where imports are over one-third of the GNP, changes in the terms of trade can have a dramatic effect on the real standard of living.

While it is easy to calculate per capita GNPs, it is notoriously difficult to make precise standard-of-living comparisons among countries. In each country, individuals naturally shift their purchases toward those items that are relatively cheap in that country. Tastes, circumstances, traditions, and habits differ. Individuals do not buy the same basket of goods and services. What is a necessity in one country may be a luxury in another. Health care may be provided by government in one country and purchased privately in another. And how do you evaluate vast expenditures, such as those we make on health care, where we are spending more than the rest of the world but getting less if you look at life expectancy (U.S. males are now sixteenth in the world)?

But whatever our precise ranking at the moment, the rest of the world is catching up, and if they have not already surpassed us, they soon will. From many perspectives, this catching-up process is desirable. Most rich people find it more comfortable to live in a neighborhood with other rich people. The tensions are less and life is more enjoyable. What is not so comfortable is the prospect that our rich neighbors will continue to grow so rapidly that we slip into relative backwardness.

Up to now, we have comforted ourselves with the belief that the economic growth of others would slow down as soon as they had caught up with us. It was simply easier to adopt existing technologies than to develop new technologies—or so we told ourselves. But as other countries have approached our productivity levels, and as individual industries in these countries have begun to be more productive, the "catching-up" hypothesis becomes less and less persuasive.

In the period from 1972 to 1978, industrial productivity rose 1 percent per year in the United States, almost 4 percent in West Germany, and over 5 percent in Japan.[5] These countries were introducing new products and improving the process of making old products faster than we were. Major American firms were reduced to marketing new consumer goods such as video recorders, which were made exclusively by the Japanese. In many industries, such as steel, we are now the ones with the "easy" task of adopting the technologies developed by others. But we don't. Instead of junking our old, obsolete open-hearth furnaces and shifting to the large oxygen furnaces and continuous casting of the Japanese, we retreat into protection against the "unfair" competition of Japanese steel companies. The result is a reduction in real incomes as we all pay more for steel than we should. As a result, our economy ends up with a weak steel industry that cannot compete and has no incentive to compete, given its protection in the U.S. market.

This relative economic decline has both economic and political impacts. Economically, Americans face a relative decline in their standard of living. How will the average American react when it becomes obvious to the casual tourist (foreigners here, Americans there) that our economy is falling behind? Since we have never had that experience, no one knows; but if we are like human beings in the rest of the world, we won't like it. No one likes seeing others able to afford things that they cannot.

As gaps in living standards grow, so does dissatisfaction with the performance of government and economy. The larger the income gap, the more revolutionary the demands for change. Today's poor countries are in turmoil, but it should be remembered that these countries are not poor compared with the poor centuries ago. They are only poor relative to what has been achieved in today's rich countries. If we become relatively poor, we are apt to be just as unhappy.

Politically a declining economy means that we have to be willing to make greater sacrifices in our personal consumption to maintain any level of world influence. This can be done. The Russians have become our military and geopolitical equals despite a per capita GNP that is much lower than ours. They simply put a larger fraction of their GNP into defense. But the need to cut consumption creates strains in a democracy that do not exist in a dictatorship. Americans may gradually decide that they cannot afford to maintain a strategic military capability to defend countries that are richer than they are. They may decide that they cannot afford to lubricate peace settlements, such as that between Israel and Egypt, with large economic gifts. Some of the international economic burdens could be shifted to our wealthier allies, but this would inevitably mean letting them make more of the important, international decisions. In many circumstances (Israel vs. Egypt?)

the Germans and the Japanese may not make the same decisions that we would make.

The hard-core conservative solution is to "liberate free enterprise," reduce social expenditures, restructure taxes to encourage saving and investment (shift the tax burden from those who save, the rich, to those who consume, the poor), and eliminate government rules and regulations that do not help business. Specifically, the capital gains taxes that were reduced in 1978 should be reduced further; the "double" taxation of dividends should be ended; income transfer payments to the poor and the elderly should be frozen; environmentalism should be seen as an economic threat and rolled back. Laffer curves sprout like weeds to show that taxes should be cut to restore personal initiative. Only by returning to the virtues of hard work and free enterprise can the economy be saved.

In thinking about this solution, it is well to remember that none of our competitors became successful by following this route. Government absorbs slightly over 30 percent of the GNP in the United States, but over 50 percent of the GNP in West Germany. Fifteen other countries collect a larger fraction of their GNP in taxes.[6]

Other governments are not only larger; they are more pervasive. In West Germany, union leaders must by law sit on corporate boards. Sweden is famous for its comprehensive welfare state. Japan is marked by a degree of central investment planning and government control that would make any good capitalist cry. Other governments own or control major firms, such as Volkswagen or Renault. Ours is not the economy with the most rules and regulations; on the contrary, it is the one with the fewest rules and regulations. As many American firms have discovered to their horror, it simply isn't possible to fire workers abroad as it is here. It is a dubious achievement, but nowhere in the world is it easier to lay off workers.

Nor have our competitors unleashed work effort and savings by increasing income differentials. Indeed, they have done exactly the opposite. If you look at the earnings gap between the top and bottom 10 percent of the population, the West Germans work hard with 36 percent less inequality than we, and the Japanese work even harder with 50 percent less inequality.[7] If income differentials encourage individual initiative, we should be full of initiative, since among industrialized countries, only the French surpass us in terms of inequality.

Moreover, our own history shows that our economic performance since the New Deal and the onset of government "interference" has been better than it was prior to the New Deal. Our best economic decades were the 1940s (real per capita GNP grew 36 percent), when the economy was run as a command (socialist) wartime economy, and the 1960s (real per capita GNP grew 30 percent), when we had all that growth in social welfare programs.[8] Real per capita growth since the advent of government intervention has been more than twice as high as it was in the days when governments did not intervene or have social welfare programs.

The British are often held up as a horrible example of what will happen to us if we do not mend our ways and reverse the trend toward big government. But whatever is wrong with the British economy, it has little to do with the size of government. British growth fell behind that of the leading industrial countries in the nineteenth century and has remained behind ever since. Slow growth did not arrive with the Labour government in 1945. On the contrary, British growth since 1945 has actually been better than before. There is no doubt that the British economy is in sad shape, but as the West Germanys of the world demonstrate, its problems are not a simple function of government size.

As both our experience and foreign experience demonstrate, there is no conflict between social expenditures or government intervention and economic success. Indeed, the lack of investment planning, worker participation, and social spending may be a cause of our poor performance. As we, and others, have shown, social reforms can be productive, as well as just, if done in the right way. If done in the wrong way, they can, of course, be both disastrous and unjust. There may also be some merit in "liberating free enterprise" if it is done in the right way. There are certainly unnecessary rules and regulations that are now strangling our economy. The trick is not rules versus no rules, but finding the right rules.

The American problem is not returning to some golden age of economic growth (there was no such golden age) but in recognizing that we have an economic structure that has never in its entire history performed as well as Japan and West Germany have performed since World War II. We are now the ones who must copy and adapt the policies and innovations that have been successful elsewhere. To retreat into our mythical past is to guarantee that our days of economic glory are over.

Unsolvable Problems

But our problems are not limited to slow growth. Throughout our society there are painful, persistent problems that are not being solved by our system of political economy. Energy, inflation, unemployment, environmental decay, ever-spreading waves of regulations, sharp income gaps between minorities and majorities—the list is almost endless. Because of our inability to solve these problems, the lament is often heard that the U.S. economy and political system have lost their ability to get things done. Meaningful compromises cannot be made, and the politics of confrontation are upon us like the plague. Programs that would improve the general welfare cannot be started because strong minorities veto them. No one has the ability to impose solutions, and no solutions command universal assent.

The problem is real, but it has not been properly diagnosed. One cannot lose an ability that one never had. What is perceived as a lost ability to act is in fact (1) a shift from international cold war problems to domestic problems, and (2) an inability to impose large economic losses explicitly.

As domestic problems rise in importance relative to international problems, action becomes increasingly difficult. International confrontations can be, and to some extent are, portrayed as situations where everyone is fairly sharing sacrifices to hold the foreign enemy in check. Since every member of society is facing a common threat, an overwhelming consensus and bipartisan approach can be achieved.

Domestic problems are much more contentious in the sense that when policies are adopted to solve domestic problems, there are *American* winners and *American* losers. Some incomes go up as a result of the solution; but others go down. Individuals do not sacrifice equally. Some gain; some lose. A program to raise the occupational position of women and minorities automatically lowers the occupational position of white men. Every black or female appointed to President Carter's cabinet is one less white male who can be appointed.

People often ask why President Kennedy was so easily able to get the Man on the Moon project underway, while both Presidents Nixon and Ford found it impossible to get their Project Independence underway. There is a very simple answer. Metaphorically, some American has to have his or her house torn down to achieve energy independence, but no American lives between the earth and the moon. Everyone is in favor of energy independence in general, but there are vigorous objectors to every particular path to energy independence. In contrast, once a consensus had been reached on going to the moon, the particular path could be left to the technicians. In domestic problems, the means are usually as contentious as the ends themselves.

As we shall see in later chapters, there *are* solutions for each of our problem areas. We do not face a world of unsolvable problems. But while there are solutions in each case, these solutions have a common characteristic. Each requires that some large group —sometimes a minority and sometimes the majority—be willing to tolerate a large reduction in their real standard of living. When the economic pluses and minuses are added up, the pluses usually exceed the minuses, but there are large economic losses. These have to be allocated to someone, and no group wants to be the group that must suffer economic losses for the general good.

Recently I was asked to address a Harvard alumni reunion on the problem of accelerating economic growth. I suggested that we were all in favor of more investment, but that the heart of the problem was deciding whose income should fall to make room for more investment. Who would they take income away from if they were given the task of raising our investment in plant and equipment from 10 to 15 percent of the GNP? One hand was quickly raised, and the suggestion was made to eliminate welfare payments. Not surprisingly, the person was suggesting that someone else's income be lowered, but I pointed out that welfare constitutes only 1.2 percent of the GNP.[9] Where were they going to get the remaining funds—3.8 percent of GNP? Whose income were they willing to cut after they had eliminated government programs for the poor? Not a hand went up.

A Zero-Sum Game

This is the heart of our fundamental problem. Our economic problems are solvable. For most of our problems there are several solutions. But all these solutions have the characteristic that someone must suffer large economic losses. No one wants to volunteer for this role, and we have a political process that is incapable of forcing anyone to shoulder this burden. Everyone wants someone else to suffer the necessary economic losses, and as a consequence none of the possible solutions can be adopted.

Basically we have created the world described in Robert Ardrey's *The Territorial Imperative*. To beat an animal of the same species on his home turf, the invader must be twice as strong as the defender. But no majority is twice as strong as the minority opposing it. Therefore we each veto the other's initiatives, but none of us has the ability to create successful initiatives ourselves.

Our political and economic structure simply isn't able to cope with an economy that has a substantial zero-sum element. A zero-sum game is any game where the losses exactly equal the winnings. All sporting events are zero-sum games. For every winner there is a loser, and winners can only exist if losers exist. What the winning gambler wins, the losing gambler must lose.

When there are large losses to be allocated, any economic decision has a large zero-sum element. The economic gains may exceed the economic losses, but the losses are so large as to negate a very substantial fraction of the gains. What is more important, the gains and losses are not allocated to the same individuals or groups. On average, society may be better off, but this average hides a large number of people who are much better off and large numbers of people who are much worse off. If you are among those who are worse off, the fact that someone else's income has risen by more than your income has fallen is of little comfort.

To protect our own income, we will fight to stop economic change from occurring or fight to prevent society from imposing the public policies that hurt us. From our perspective they are not good public policies even if they do result in a larger GNP. We want a solution to the problem, say the problem of energy, that does not reduce our income, but all solutions reduce someone's income. If the government chooses some policy option that does not lower our income, it will have made a supporter out of us, but it will have made an opponent out of someone else, since someone else will now have to shoulder the burden of large income reductions.

The problem with zero-sum games is that the essence of problem solving is loss allocation. But this is precisely what our political process is least capable of doing. When there are economic gains to be allocated, our political process can allocate them. When there are large economic losses to be allocated, our political process is paralyzed. And with political paralysis comes economic paralysis.

The importance of economic losers has also been magnified by a change in the political structure. In the past, political and economic power was distributed in such a way that substantial economic losses could be imposed on parts of the population if the establishment decided that it was in the general interest. Economic losses were allocated to particular powerless groups rather than spread across the population. These groups are no longer willing to accept losses and are able to raise substantially the costs for those who wish to impose losses upon them.

There are a number of reasons for this change. Vietnam and the subsequent political scandals clearly lessened the population's willingness to accept their nominal leader's judgments that some project was in their general interest. With the civil rights, poverty, black power, and women's liberation movements, many of the groups that have in the past absorbed economic losses have become militant. They are no longer willing to accept losses without a political fight. The success of their militancy and civil disobedience sets an example that spreads to other groups representing the environment, neighborhoods, and regions.

All minority groups have gone through a learning process. They have discovered that it is relatively easy with our legal system and a little militancy to delay anything for a very long period of time. To be able to delay a program is often to be able to kill it. Legal and administrative costs rise, but the delays and uncertainties are even more important. When the costs of delays and uncertainties are added into their calculations, both government and private industry often find that it pays to cancel projects that would otherwise be profitable. Costs are simply higher than benefits.

In one major environmental group, delays are such a major part of their strategy that they have a name for it—analysis paralysis. Laws are to be passed so that every project must meet a host of complicated time-consuming requirements. The idea is not to learn more about the costs and benefits of projects, but to kill them. If such requirements were to be useful in deciding whether a project should be undertaken, environmental-impact statements, for example, would have to be inexpensive, simple, and quick to complete. Then a firm might undertake the studies to help determine whether they should or should not start a project.

Instead, the studies are to be expensive and complex to serve as a financial deterrent to undertaking any project, to substantially lengthen the time necessary to complete any project, and to ensure that they can be challenged in court (another lengthy process). As a consequence, the developer will start the process only if he has already decided on other grounds to go ahead with the project. The result is an adversary situation where the developer cannot get his project underway—and where the environmentalists also cannot get existing plants (such as Reserve Mining) to clean up their current pollution. Where it helps them, both sides have learned the fine art of delay.

Consider the interstate highway system. Whatever one believes about the merits of completing the remaining intracity portion of the system, it is clear that it gives the country an intercity transportation network that would be sorely missed had it not been built. Even those who argue against it do so on the grounds that if it had not been built, some better (nonauto) system would have been devised. Yet most observers would agree that the interstate highway system could not have been built if it had been proposed in the mid-1970s rather than in the mid-1950s.

Exactly the same factors that would prevent the initiation of an interstate highway system would also prevent the initiation of any alternative transportation system. A few years ago, when a high-speed rail system was being considered for the Boston-Washington corridor, a former governor of Connecticut announced that he would veto any relocation of the Boston-to-New York line on the grounds that it would be of prime benefit to those at either end of the line, but would tear up Connecticut homes. The groups opposing an intercity rail network would be slightly different from the groups opposing an intercity highway network, but they would be no less effective in stopping the project. Any transportation system demands that land be taken and homes be torn down. At one time, this was possible; at the moment, it is impossible.

The Balkanization of nations is a worldwide phenomenon that the United States has not escaped. Regions and localities are less and less willing to incur costs that will primarily help people in other parts of the same country. Consider the development of the coalfields of Wyoming and Montana. There is no question that most of the benefits will accrue to those living in urban areas in the rest of the country while most of the costs will be imposed on those living in that region. As a result, the local population objects. More coal mining might be good for the United States, but it will be bad for local constituents. Therefore they will impose as many delays and uncertainties as possible.

The same problem is visible in the location of nuclear power plants. Whatever one believes about the benefits of nuclear power, it is clear that lengthy delays in approving sites serve no purpose other than as a strategy for killing the projects. If the projects are undertaken anyway, the consumer will have to suffer the same risks and pay the higher costs associated with these delays. What is wanted is a quick yes or no answer; but this is just what we find impossible to do. The question of nuclear power sites also raises the Balkanization issue. Whatever the probabilities of accidents, the consequences of such failures are much less if the plants are located in remote areas. But those who live in remote areas do not want the plants, since they suffer all the potential hazards and do not need the project. Everyone wants power, but no one wants a power plant next to his own home.

Domestic problems also tend to have a much longer time horizon. In modern times, even long wars are won or lost in relatively short periods of time. In contrast, a project such as energy independence would take decades to achieve. The patience and foresight necessary for long-range plans is generally not an American virtue. Consequently, representatives seeking reelection every two, four, or six years want to support programs that will bring them votes. They do not want to stick their necks out for a good cause that may conflict with their careers. Even more fundamentally, domestic problems often involve long periods where costs accrue, with the benefits following much later. Think about energy independence. For a long time, sacrifices must be made to construct the necessary mines and plants. Benefits emerge only near the end of the process. The politician who must incur the costs (raise the necessary revenue and incur the anger of those who are hurt as the projects are constructed) is unlikely to be around to collect the credits when energy independence has been achieved.

The Retreat to Government

Given the problem of loss allocation, it is not surprising that government stands in the middle of an adversary relationship. Each group wants government to use its power to protect it and to force others to do what is in the general interest. Energy producers want prices to go up and the real income of energy consumers to go down. Energy consumers want prices to go down and a reduction in the income of producers. Each understands that the government could stop them from having to suffer such losses. Each of us demands what collectively is impossible. But as the demands for protection grow, the basic assumptions of the democratic process are undermined.

To be workable, a democracy assumes that public decisions are made in a framework where there is a substantial majority of concerned but disinterested citizens who will prevent policies from being shaped by those with direct economic self-interests. Decisions in the interests of the general welfare are supposed to be produced by those concerned but disinterested citizens. They are to arbitrate and judge the disputes of the interested parties. As government grows, however, the number of such citizens shrinks. Almost everyone now has a direct economic stake in what government does in an area such as energy.

The Watergate and associated corporate bribery scandals revealed the illegal side of this problem, but the real problem is not so much illegal acts as it is the incentive to use legal ones. With everyone's economic self-interest at stake, we all form perfectly proper lobbying groups to bend decisions in our favor. But with the disinterested citizen in a minority, how are decisions to reflect the general welfare? Who is to arbitrate? Our natural inclination is to rely on the adversary process, where different self-interested groups present their case. But somewhere there has to be a disinterested judge with the power to decide or tip a political decision in the right way. The general welfare is not always on the side of those who can mobilize the most economic and political power in their own behalf. If we really were to enforce the rule that no one could vote on an issue if his or her income would go up or down as a result of the action, we would end up with few or no voters on most issues. The problem is to establish a modicum of speedy, disinterested decision-making capacity in a political process where everyone has a direct self-interest.

The Need for Distribution Judgments

We have a governmental process that goes to great lengths to avoid having to overtly lower someone's income. Such decisions are always being made of course, but they are made implicitly, under the guise of accomplishing other objectives. Conservatives are now arguing for a restructuring of taxes and expenditures that would make the distribution of income more unequal, but they do not defend this goal overtly. Inequality is simply a regrettable necessity on the way to higher growth.

Fortunately or unfortunately we have reached a point where it is no longer possible to solve our economic problems and still make such implicit distributional decisions. The problems are stark enough and the options clear enough that everyone both knows and cares about the distributional consequences that will follow. Deregulating the price of energy might be the efficient thing to do, but we have great trouble doing so. Everyone whose income will go down knows it, objects, and stands ready to fight the proposal.

Since government must alter the distribution of income if it is to solve our economic problems, we have to have a government that is capable of making equity decisions. Whose income ought to go up and whose income ought to go down? To do this, however, we need to know what is equitable. What is a fair or just distribution of economic resources? What is a fair or just procedure for distributing income? Unless we can specify what is equitable, we cannot say whose income ought to go down. Unless we can say whose income ought to go down, we cannot solve our economic problems.

The difficulties of specifying economic equity neither obviate the need for equity decisions nor stop such decisions from being made. Every time a tax is levied or repealed, every time public expenditures are expanded or contracted, every time regulations are extended or abolished, an equity decision has to be made. Since economic gains are relatively easy to allocate, the basic problem comes down to one of allocating economic losses. Whose income "ought" to go down?

Historically we have used economic growth to avoid having to make this judgment. If we just have more growth, we can have more good jobs for everyone, and we won't have to worry about taking jobs away from whites and giving them to blacks. If we just have more economic growth, we won't have to worry about government collecting taxes in the Northeast and spending them in the Southwest. More is obviously better than less, and economic growth has been seen as the social lubricant that can keep different groups working together.

American liberals and conservatives both used to regard economic growth as unambiguously good. Through the magic of economic growth and individual self-interest, everyone would have more. If everyone had a higher income, then society would not have to address the divisive issue of equity or what constitutes a just distribution of economic resources. Individuals would be happy with their new, higher incomes regardless of their relative status.

We now know that almost all the implicit assumptions in this social consensus are false. When we are talking about incomes above the range of psychological necessities, individual perceptions of the adequacy of their economic performance depend almost solely on relative as opposed to absolute position. The poor in the United States might be rich in India, but they actually live in the United States and feel poor. The middle class may have fresh fruits and vegetables that the richest kings could not afford in the Middle Ages, but they feel deprived relative to the upper-middle class, who can afford things they cannot afford.

The proportion of any population that report themselves satisfied with their economic performance rises not at all as that population's average income rises.[10] Those happiest with their economic circumstances have above-average incomes. There is no minimum absolute standard of living that will make people content. Individual wants are not satiated as incomes rise, and individuals do not become more willing to transfer some of their resources to the poor as they grow richer. If their income rises less rapidly than someone else's, or less rapidly than they expect, they may even feel poorer as their incomes rise.

This immediately forces a democracy into a no-win situation where, whatever it decides about the just distribution of resources, there will be a large number (perhaps even a majority) of unhappy voters. Distributional issues are highly contentious and precisely the kind of issues that democracies find it most difficult to solve. It is not *we* versus *them,* but *us* versus *us* in a zero-sum game.

In the past there was also a widespread optimistic belief that the distribution of market incomes would automatically become more equal with growth. Minorities would automatically catch up with majorities, and the poor would close the gap between themselves and the rich. Except for the physically handicapped, everyone would be able to reach an acceptable minimum standard of living if output were only high enough. And for those few who could not earn their own way, the richer our society, the easier it would be to give them an adequate income.

Here again, we know now that this is not the case. From 1948 to 1978 the distribution of earnings grew more unequal. In 1939 the average year-round, full-time female worker earned just 61 percent of what the equivalent male made. By 1977 she made just 57 percent as much.[11] Forty years of rapid economic growth, yet women were farther behind at the end than at the beginning.

The Drive for Economic Security

All this is exacerbated by an increasing drive for economic security. Economic security is to modern man what a castle and a moat were to medieval man. One would have expected that the desire for economic security would fall as the danger of real starvation and exposure faded into the past. But this hasn't happened. Instead, the desire for economic security is probably the major economic demand confronting the political marketplace. Everyone wants economic security, and government is seen as the prime vehicle for guaranteeing it. The drive for economic security dominates our actions and may end up dominating our economy.

The desire for economic security can be seen in both how we earn our incomes and how we spend them. When public opinion polls ask about desired job characteristics, economic security always takes top place—well above higher pay.[12] This preference for economic security shows up in many ways. Old workers want seniority hiring and firing so that worries about layoffs can be confined to someone else—new workers. Restrictive work rules are designed to provide job security.

As for consumers, "let the buyer beware" is not an aphorism that attracts much support nowadays. Yet with better-educated buyers who make fewer mistakes and have higher incomes, so that they can afford mistakes more easily, we should be moving in the direction of letting individuals make more of their own decisions. But we aren't.

We are much less willing to let individuals make their own mistakes. Anyone who has bought a house under a federally insured mortgage knows that the regulations act as if the buyer were a first-class idiot. Consumer legislation usually assumes that consumers are incompetent. It is common to explain these regulations as having been forced upon society by some extremely powerful minority that wants to torment the current economic system. This is a mistake. The problem is to understand why most of us want to be protected from our own mistakes.

Everyone wants economic security and runs to the government for protection when he feels it slipping away. When OPEC raised the price of oil in 1973–74 and food prices exploded, both were met with overwhelming demands for government regulations to mitigate the real income losses. Energy became a regulated industry and export embargoes were imposed on grain sales. Examples are endless: farmers, the elderly, the steel industry, electronics, textiles—everyone wants economic security. None of the groups are villains. They simply want what each of us wants—economic security.

Some of the demand for security springs from the nature of industrial societies. In agricultural societies, economic destruction was seen as, and mainly was, the result of impersonal, uncontrollable forces: if the weather is bad, incomes are going to fall and no earthly force can alter the results. Economic destruction in industrial societies is caused by identifiable human actions that can be controlled. If someone plans to build a coal-slurry pipeline from the coalfields of Wyoming to the Midwest, the income of railroaders will fall, but they can mobilize to prevent the pipeline companies from getting the right of eminent domain necessary to build the pipeline. If incomes are threatened by Japanese steel or TV sets, Japanese products can be identified and kept out. In an industrial society, economic security becomes a feasible objective.

Modern industrial societies may also lead to a set of financial interrelationships (mortgages, consumer credit, pension rights, and so forth) in which small declines in personal income are more threatening to an individual than they were in the past. Objectively, being hungry may be more serious than having your car repossessed, but subjectively the repossession may pose more of a threat to modes of living. With income security during retirement depending upon private pensions, job security while working becomes directly tied to income security during one's old age. Industrial sons and daughters are not expected to take care of industrial mothers and fathers in their old age. Instead they depend on pensions that are attached to jobs.

Skills are also threatened in a world where many skills are learned on the job and where job openings are awarded based on seniority. To move from one employer to another involuntarily is to go to the bottom of the skills ladder and start over. To lose one's job is to take a chance on destroying one's human capital and substantially reducing one's earnings.

Economic security also has a peculiar dynamic. Every instance of providing economic security leads to demands for more economic security. If the steel industry is protected from its own inefficiencies, why shouldn't everyone else be protected from their own inefficiencies? Even more important, U.S. steel users will now have to buy steel at a higher price than their foreign competitors. This makes them less competitive and increases the probability that they will also have to ask for protection. They must have protection to offset the effects of the protection given to someone else.

As protection grows, there is no natural stopping point. The more protection we have, the more we need. Protected industries almost never reach the point where they can throw off their protection and reenter the competitive marketplace. Instead they drag others down with them.

The growth of large economic institutions also forces government to take many protective actions. At the heart of capitalism and competitive markets lies the doctrine of failure. The inefficient are to be driven out of business by the efficient. But governments cannot tolerate the failure of large economic actors. Neither the Lockheed Corporation nor New York City can be allowed to fail, since the disruptions to our integrated economy would be too large to tolerate. Needed military goods would not be delivered, and millions of bondholders would lose a substantial part of their wealth. In both cases, the rescue was organized by a conservative, free-market Republican government. Any other government would have done the same.

But if we rescue large economic actors, this creates a demand for rescuing the local grocery store or the small town from its mistakes. Unless we do so, we have a double standard for the large and the small when it comes to failure. But to rescue is to control. It is also to undercut the whole doctrine of competitive capitalism. Those who fail won't be punished economically.

If we could simply buy economic security, as one buys an insurance policy on one's life or house, economic security would not be a difficult problem. But we cannot. Instead, we give people economic security by guaranteeing them that their current earnings opportunities will not disappear. But this guarantee locks us into current activities and makes it difficult to shift to the new products and processes that are the heart of economic progress. The problem is to combine economic progress with economic security when to a great extent they are mutually incompatible. Everyone wants both, but everyone cannot have both.

Economic progress always tends to be thought of in terms of bright new products and processes, but we forget that every new product replaces some old product and every new process replaces some old process. Economic construction is based on economic destruction. In the process of destroying old products and old processes, some Americans will suffer large economic losses even though other Americans will make even larger economic gains. Only very seldom is economic growth a process without losers. Average real standards of living rise, but this gain obscures many losses.

Losers naturally want to eliminate their losses, but this can only be done by stopping the economic progress that threatens to cause their losses. As each of us, individually and in groups, searches for economic security, we collectively reduce the rate of real growth and produce an ossified society that is incapable of adjusting to new circumstances.

One simplistic solution is to give up on either economic progress or economic security. There are advocates of both positions. Conservatives, from what they believe are impregnable economic positions, generally recommend that everyone else should give up on economic security and live in a rugged, dynamic, competitive, free enterprise economy. In practice, those same people run to the government for protection if they see their own incomes threatened. None of the industries that are now protected would have been protected if the managers of those industries had not wanted protection.

Others with equally advantageous positions recommend giving up on economic growth under the cover of environmental quality or natural-resource exhaustion. The economy is to be frozen so that they can enjoy their current advantageous positions without fear of competition for the indefinite future. As they say in Colorado, a conservationist is a person who built his mountain cabin last year, while a developer is someone who wants to build his mountain cabin this year.

In practice, neither of these solutions is a solution. Too many people are not satisfied with what they have; they want more. They are not about to turn the economy off and freeze themselves into a position where there is no hope of economic advancement. Similarly, too many people want economic security. Each of us, when threatened, wants security. And in a democracy, we will organize to get what we want. The obvious goal is to deliver economic security without stopping economic progress. But how?

The demand for economic security also heightens the tension between our ideology of individual decision making and the practical necessity of collective decision making. No individual can guarantee himself economic security. Economic security is only possible if some other individual, or group of individuals, agrees to share income with you under some specified set of circumstances. By its very nature, economic security is a collective action requiring collective decisions and collective coercion. Those who make television sets can only have their income protected if the rest of us are forced to buy U.S.-made TV sets. In an economy with only individual decisions, there is no individual economic security.

Yet each of us is inconsistent. When collective coercion is used to raise our real standard of living, we are in favor of it. When it is used to limit our actions and raise someone else's income, we are against it. The same utility executive who preaches the virtues of individual enterprise objects when a neighborhood organizes to stop the construction of his power plant. Yet they are just practicing what he has been preaching.

The drive for economic security is difficult to accommodate in our mixed capitalistic economy. As we deliver economic security, we undercut the implicit assumptions of capitalism, democracy, and individual initiative. Economic failure won't hurt, because failures will be protected by government. This both reduces the rate of economic progress and removes the rationale for having capitalism in the first place. If government protects and controls, it might just as well own.

With everyone being protected, there are no concerned, disinterested citizens to make the democratic process work. Special-interest lobbies dominate, and we all belong to some special interest. The ability to decide collapses into lengthy adversary procedures where everyone is worn out and no one is the long-run winner. Costs rise, new projects cannot be undertaken, and old projects cannot be transformed. Each of us pays verbal homage on the fourth of July to individual initiative, but we run to the government whenever we are threatened.

Paralysis

At the end of the 1970s our political economy seems paralyzed. The economy is stagnant, with a high level of inflation and unemployment. Fundamental problems, such as the energy crisis, exist but cannot be solved. We have lost the ability to get things done. A successful man-on-the-moon project could be launched in the 1960s, but in the 1970s energy independence is beyond our reach.

Lacking a consensus on whose income ought to go down, or even the recognition that this is at the heart of the problem, we are paralyzed. We dislike the current situation, we wish to do something about our problems, but we endure them because we have not learned to play an economic game with a substantial zero-sum element.

But as in all boilers, when the steam rises it reaches a limit. Will the operator develop a good control valve that solves the necessary problems, or will the pressure, in time, become so intense that the lid blows off? Or have we simply reached the point where some future economic historian will say that our day in the economic sun has come to an end?

Chapter 2

Energy

NOWHERE is the nature of our fundamental dilemma more clearly illustrated than in energy. High prices, shortages, and supply disruptions are serious. They threaten future growth in our standard of living and are the main driving force behind an accelerating rate of inflation. They disrupt and disturb our lives in countless ways. At the same time, we have been unable to solve the problem. President Nixon could announce a Project Independence, and succeeding presidents could announce that the energy problem was the "moral equivalent of war," but almost a decade later we are farther from energy independence than we were at the beginning. This has occurred in spite of the fact that we are a country rich in energy resources.

The lack of action does not spring from a lack of solutions, but from the fact that each solution would cause a large, real income decline for some segment of the population. Everyone is in favor of energy independence in the abstract, but each path to energy independence is vigorously opposed by some significant group that would suffer large income declines if this particular solution were chosen. In the process, all solutions are vetoed and we remain paralyzed. The status quo is painful, but we cannot move.

Energy

Origins of the Problem

The history of our current problems began in 1957 when President Eisenhower imposed an oil import quota to protect domestic oil producers from cheap Middle Eastern oil. With the development of Middle Eastern oil fields, prices were threatening to fall and force many U.S. producers out of business. Their high-cost wells could not compete with the low-cost wells of the Middle East. But higher prices for producers means lower incomes for consumers. Whenever government policies raise someone's income, they must lower someone else's income.

The import quotas imposed by the Eisenhower Administration were designed to raise the income of oil producers and were defended on grounds of national security. They were preserving a domestic oil industry in case of war. Yet from a national security point of view, they were completely counterproductive. If defense were the real aim, government programs should have been encouraging the consumption of foreign oil during peacetime to save domestic oil for future wars. Our current military vulnerability to oil cutoffs partially springs from this 1950s so-called defense policy. If it had not been in place, we would have imported more oil and now have larger reserves of domestic oil left.

In the 1950s even keeping out foreign oil was not enough to give producers the income they wanted. Without further controls, U.S. production would have substantially exceeded U.S. consumption with the inevitable falling prices and incomes for producers. Here again government regulations were used to stop consumer incomes from rising. The Texas Railroad Commission simply limited the amount of oil that could be pumped. This policy was defended on the grounds of long-run conservation. If the fields were pumped more slowly, they would produce more oil. The feebleness of this argument can be seen from the fact that the policy was quickly abandoned when prices rose two decades later.

Thus we start with a long period where government policies were actively used to raise producer's incomes and lower consumer's incomes. At the moment, the oil industry is in favor of price deregulation and a free market, but it is well to remember that this is not their basic ideological position. The oil industry, like everyone else, is only in favor of free markets when it is in their economic self-interest to be in favor of free markets.

But as time passed, oil consumption grew faster than potential production. Where we could once produce much more than we needed, we eventually reached a point where we needed more than we could produce. By 1978 almost 50 percent of our oil was being imported.[1] While oil accounts for only 40 percent of our total energy consumption, imported oil supplies were our marginal source of energy necessary for economic growth. As demands for energy rose, these demands were being met with imported oil.

By 1973 U.S. consumption exceeded U.S. production, but the world still had production capabilities significantly larger than consumption demands. This was only true, however, if one included that massive pool of cheap Middle Eastern oil. Leaving it aside, world consumption significantly exceeded world production. As a result, any group controlling Middle Eastern oil had a significant monopoly position. By raising or lowering Middle Eastern production, they could control world oil prices. When economic opportunities exist to make massive amounts of money, it is surprising if someone does not take advantage of them, and in this case there were no surprises. OPEC was formed and tripled the price of oil in 1973–74.

If a free market in energy had been allowed to exist, this would have meant a tripling in the price of both domestic oil and other forms of energy. Buyers would have attempted to shift toward cheaper forms of domestic energy when the price of imported oil went up, but the net result would drive up the price of domestic energy to world levels. Even at much higher prices, the United States was no longer self-sufficient.

The Free Market Solution

One solution to the energy problem is simply to let the price of energy rise in accordance with that of imported oil. This would solve the problem in the sense that there would be no gas lines, no shortages, and no Energy Department full of complex and sometimes counterproductive regulations. Supply and demand pricing would work, but at the same time it would involve an enormous change in the distribution of income.

From the perspective of Americans, higher oil prices represent a mixture of gains and losses. For the country as a whole, Americans lose since they had to pay an extra $30 billion (in 1978) for imported oil over and above what they would have had to pay if prices had stayed at 1972 levels.[2] Since $30 billion represents about 1.5 percent of our GNP, average real incomes have to fall by about that amount. But domestic energy producers also gain if domestic energy prices are allowed to rise. Since about 80 percent of our energy is domestically produced, this transfer between Americans from energy consumers to energy producers is much larger than the transfer to foreigners. In 1978 an additional $120 billion or 6 percent of the GNP would have been transferred from American consumers to American producers if all energy prices had been allowed to follow the price of imported oil.[3] On the average, Americans are not poorer because of this transfer–it is from one American to another–but particular Americans will experience large income gains and other Americans will experience large income losses. Given an expected increase in the price of imported oil of about 100 percent during 1979, the 1979 transfers would be correspondingly larger. No one willingly accepts such a reduction in their income.

The distribution of losses depends upon who consumes energy (see table 2–1). Each of us is a direct consumer of energy in the form of gasoline and home heating or cooling, but each of us is also an indirect consumer of energy in the products we buy. Every product embodies energy in its production and distribution.

As you can see in table 2–1, the proportion of income going to energy consumption differs dramatically between the rich and the poor and less dramatically across regions. While a 100 percent increase in the price of energy would reduce the real income of the average American by 9.9 percent, it would have reduced the real income of the poorest decile of families by 34 percent and the richest decile by 5 percent.[4] The real income effects among the poor are almost seven times as large as they are among the rich.

TABLE 2–1
Direct Household Energy Consumption as a Percentage of Before-Tax Income

Decile * (Poorest to Richest)	Home Energy Consumption (%)	Gasoline Consumption(%)	Indirect Consumption (%)	Total (%)
First	20.2	9.6	4.3	34.1
Second	10.4	5.8	3.9	20.1
Third	7.4	5.6	3.5	16.5
Fourth	5.6	5.2	3.1	13.9
Fifth	4.7	4.8	2.7	12.2
Sixth	3.9	4.5	2.3	10.7
Seventh	3.9	3.8	2.0	9.7
Eighth	3.3	3.7	1.7	8.7
Ninth	3.0	3.1	3.1	7.4
Tenth	2.0	2.2	1.0	5.2
TOTAL:	3.8	3.6	2.5	9.9

* Decile = 10% of population.

TABLE 2–2
Direct Household Energy Consumption by Region

Region	Home Energy Consumption (%)	Gasoline Consumption (%)	Indirect Consumption (%)	Total (%)
Northeast	4.5	3.3	2.5	11.4
Northcentral	3.3	3.7	2.5	9.5
South	3.9	4.0	2.5	10.4
West	2.3	3.7	2.5	8.5

SOURCE (for Table 2–1 and Table 2–2): Estimated from data in U.S. Department of Labor, *Consumer Expenditure Survey Series 1973–74,* and input-output tables of U.S. Department of Labor.

Between regions the real income decline in the Northeast (11.4 percent) is about 30 percent larger than that in the West (8.5 percent). (It should of course be remembered that these energy costs include the cost of delivering and distributing energy. A 100 percent increase in the price of crude energy does not lead to a 100 percent increase in the price of delivered energy.)

While it is relatively easy to calculate whose income would go down as energy prices rise, it is much harder to calculate whose income would go up. The income of those who own energy resources would go up, but who are they? No one knows with any certainty, but it is possible to find an approximate answer. Since most of our energy resources are owned by corporations, the ownership of energy is probably very similar to the ownership of corporate stock. Here we know that the top 10 percent of all households owns over 90 percent of all corporate stock.[5]

If we assume the same situation is true with respect to energy resources (the top 10 percent owns 90 percent), most of the income transfers among Americans will go to the top 10 percent of the population. When the pluses and minuses are added up, their income will go up, and the income of the remaining 90 percent will go down. Since the income gains to the top 10 percent from owning energy resources would be about five times as large as their income losses from having to pay higher energy prices, a free market for energy would have resulted in a sharp shift toward inequality in the distribution of income.

Given that the free market solution leads to large income losses for a large fraction of the country, it is not surprising in a democracy that government stepped in to prevent these losses from occurring. The U.S. government could not stop the $30 billion transfer to foreigners, but it could and did stop much of the $120 billion transfer among Americans.

Spreading Waves of Regulation

Domestic energy prices were frozen in 1973–74, but this policy created a set of circumstances that required a vast elaboration of controls. First, imported oil is not consumed equally across the country. Some regions use mostly imported oil; some regions use only domestic oil. Since imported oil prices cannot be controlled, actions have to be taken to equalize oil prices if those regions that use imported oil (the Northeast) are not to suffer large income losses while others suffer not at all.

To equalize prices a complicated set of taxes and subsidies was initiated, but the essence of the system is simple. Suppose that 50 percent of our oil is imported and sells for $30 per barrel, and 50 percent of our oil is domestically produced and held at a price of $10 per barrel. Cheap domestic oil and expensive foreign oil are mixed together and sold at the average price of the two—in this case $20 [(.5) ($30) + (.5) ($10)]. A tax of $10 per barrel is levied on domestic oil production and the revenue ($10 per barrel) is used to subsidize oil importers. With these taxes and subsidies, both importers and domestic producers can profitably sell oil for $20 per barrel.

This regulation means that everyone in the country faces the same price for oil, but it in turn creates two new problems. Producers do not have an incentive to look for new energy resources, and consumers do not have an incentive to conserve. The production problem arises because you can make as much or more profits importing oil as you can producing domestic oil with price equalization. You might just as well save your domestic oil until some future date when prices are higher and controls have been lifted. To offset this effect some modification of the controls must be made to restore incentives for more domestic energy production.

To encourage production the regulations differentiate among different oil sources. Oil for new oil fields can be sold at world prices. Oil from new wells in old fields can be sold at a price higher than that from old wells in old fields, but lower than that from new wells in new fields. Different prices are set for different secondary recovery methods (pumping steam down to force more oil up, and so forth). The result is a proliferation of prices with many types of oil and many opportunities to make oil look "newer" than it actually is. Regulations become cumbersome and complex, and more profits were to be made by the skillful political manipulator than by the efficient oil producer.

The conservation problem arises since our marginal source of oil costs $30 but oil is sold for an average price of $20. Consumers use the lower price to decide how much oil to buy, but in doing so they use a price that is below our real costs. De facto, we are subsidizing the consumption of imported oil. To counterbalance this effect we now adopt a whole series of regulations designed to conserve fuel. Mileage standards and speed limits are set for cars, maximum and minimum temperatures are set for commercial buildings, and tax credits are given for insulation. Each regulation is complex, and all of them are less efficient than $30 oil in discouraging the use of energy.

Pricing energy at world levels will eventually reduce consumption and increase production, but there is a major political problem. The enormous income transfer starts immediately, but the higher production and lower consumption only occur slowly and gradually. The costs are now while the benefits lie in the future.

In the short run, each of us is locked into a pattern of energy consumption. We own a certain car, live a certain distance from work, and heat and cool a home with certain energy properties. As a result, a 100 percent increase in the price of energy only lowers consumption by 10 percent in the short run.[6] As time passes, however, we adjust more and more of our energy patterns. The new car gets better gas mileage, commuting distances can be shortened, and homes can be made more energy efficient. Countries where energy prices have always been high produce their economic goods and services with much less energy (often half as much). We can do the same, but it takes a long time to get from here to there. Today's effects are always small, yet if we do not start on the path to less energy consumption through higher prices, we will never arrive at the desired results.

The same problem exists on the production side of the problem. In the short run, higher prices are going to give you very little extra energy. It takes time to build new production facilities. In the long run, shifts toward other energy sources, such as coal, can be very significant.

Attempts to avoid market pricing also lead to ever-widening waves of regulations. Regulations can cope with the situation for a period of time (they have been in place since 1973–74), but they cannot cope indefinitely. They are too cumbersome to adjust to the inevitable shocks, such as the Iranian revolution, that will occur. Shortages and gas lines appear with their disruptions of both work and pleasure. Prices rise because truckers are spending time looking for fuel.

The regulatory approach, however, suffers from a more fatal flaw than inconvenience. Government regulations can control prices and to a lesser extent production, but in our system they cannot control new investments. No one can be forced to invest. Eventually new investments are necessary, and they will not be made unless they are as profitable as investments made in countries that do not control energy prices. This leads to increasingly severe shortages, as the necessary new facilities are not built to accommodate rising demands and new products (unleaded gas). Eventually we are forced to decide whether we want free market pricing (with its large income losses and gains) or a nationalized energy industry where government makes the necessary new investments.

The free market option is opposed by those whose income would go down. The nationalization option is opposed by those who are ideologically against government ownership, but it does not avoid the income-distribution question either. If large government investments are to be made, the funds to make these investments must be taken away from someone. Who? No one wants to lower their standard of living to make way for massive public investments any more than they want to contribute to the income of private energy producers.

Fortunately or unfortunately, government regulation is not a good halfway point between the free market and government ownership. In the long run either the free market or government ownership will dominate regulated private industry. The vice of moderation is also seen in the suggestions that we deregulate the price of energy but attempt to mitigate the income effects with energy stamps for the poor and selective tax cuts for the middle class. This compromise simply sharpens the distributional issues. Whose taxes should go up to pay for energy stamps and tax cuts for the middle class?

If all of the revenue is taken away from oil producers (windfall profit taxes), we are back to where we started, and producers have no incentive to make the necessary investments. If we take just some of the extra revenue away from oil producers, then there is only enough revenue to mitigate some of the income shocks. Among those who are going to suffer real income losses, who should get protection when protection cannot be given to everyone? Taxes could be raised to take some of the necessary revenue away from other groups, but what groups? There certainly are no volunteers.

Consumers want a full offset; producers want no offset. Both may hold political veto powers and be able to prevent the other's desires from being legislated. The current system does not work, but it is impossible to shift to a different system. Attempting to find solutions to the energy crisis without imposing large income losses is like wandering ever deeper into a spider's web. The farther you go, the more entangled you get. The problem becomes more severe and the solutions more distant.

The essence of the dilemma can be seen in the price of gasoline. In every other industrialized country, gasoline is not only sold at the full cost of imported oil (about $1 per gallon in 1978) but corporations are forced to pay an excise tax that raises the consumer's price into the $2 to $3 per gallon range. Why is it that everyone else thinks that very high gas prices are an important part of their energy policy? The rest of the world thinks that the economic burdens of paying for imported oil and the military problems of being dependent upon the volatile Middle East demand strong policies to discourage driving. Our policies encourage driving. Is everyone else wrong? Or are we simply unable to do what we should be doing?

In this case, the problem is not even the need to lower the real income of drivers, but merely to appear to be lowering the real income of drivers. Any tax revenue collected with a gasoline tax could be rebated in lower income taxes. Drivers are the numerical majority everywhere, but other countries set out to penalize driving. In contrast, our government is paralyzed at the mere appearance of lowering someone's income.

Energy Independence

The same problem of economic losses can be seen in the political stalemate over energy independence. While many countries cannot become energy self-sufficient since they lack the necessary resources, the United States could easily achieve energy independence from a technical perspective. Our coal reserves alone are so enormous that they could fill all of our needs for the foreseeable future. Many other forms of energy—nuclear, small-scale hydroelectric, solar, wind—could be harnessed to provide some of our energy needs. Technologically energy independence is well within our reach. Politically it may not be within our reach at all. Every path to energy independence requires a sharp reduction in the income of some group, and as a result every path to energy independence has its political foes. Even if these foes cannot muster the votes to cleanly defeat a particular path, they can raise costs and delay the projects so long that the path loses its appeal.

If the problem were simply one of high-priced energy due to natural scarcities, energy independence would not be a desirable goal economically. Efficiency would call for selling those goods that we are most capable of producing and buying those goods that we are least capable of producing. If energy fell into the import category, rather than the export category, there would be no cause for concern. The basic current problem in energy, however, is not scarcity but a cartel that controls the marginal source of energy.

This means that high prices are compounded by uncertain supplies. Unexpected disruptions occur because of political events in other countries, and supplies may suddenly be cut off in an attempt to force the United States to alter its foreign policies. In this context energy independence becomes a sensible goal when it otherwise would not be. It may even be sensible if it means using energy that costs more than imported oil.

While there are many energy alternatives in the United States, there are also many fundamental stumbling blocks. The first issue is the one of costs. The costs of all alternative energy sources are highly dependent upon the regulatory environment in which they are to be used. Coal is cheap in a world where no one cares about the environmental damage occurring when it is mined or burned, but coal can be a very expensive alternative energy source in an economy with strict pollution controls. This means that there is a direct confrontation between those who want cheap goods and the cheap energy necessary to produce them, and those who want a clean environment.[7] As we shall see in chapter 5, these are different individuals in different economic classes.

Similarly, nuclear energy is cheap or expensive depending upon the risks that one is willing to take with exposure to radiation. If one wants to reduce the risks close to zero, nuclear energy is very expensive; if one is willing to take substantial risks, nuclear energy is cheap. Individuals differ on the risks they are willing to take, but our willingness to tolerate nuclear accidents also has something to do with how close each of us lives to a nuclear power plant.

The regulatory environment is a major part of the costs of alternative energy sources, but the conflicts about what constitutes the "right" regulatory environment are even more costly. Changes in the health, safety, or environmental standards when a plant is half-built or fully-built, are much more expensive than any set of stable requirements known before construction begins. But not having arrived at an agreed upon set of environmental standards, no one can promise a stable set of requirements. The result is great uncertainty with lengthy time delays as we fight over what environmental standards should be imposed.

In addition to the conflicts between those who want more conventional goods and services and those who want a clean, healthy environment, each alternative energy source has a set distributional conflict of its own. Consider our most likely candidate for energy independence–Western coal. While coal is found in abundance in Wyoming and Montana, water is not. Yet coal gasification requires water. This means that the coal must be brought to water or water to coal. What current user is willing to give up the necessary water in the arid West? Who is willing to stop earning their living and give up their water to make energy for the needs of the country? No one.

Suppose that the coal is brought east to the necessary water supplies. Millions of tons of coal gasified or burned in electrical generating plants mean millions of tons of residue or fly ash that must be dumped. Where? Whose land is going to be used as the necessary slag heaps? We could transport millions of tons of coal east and then transport millions of tons of fly ash west to dump in the pits where the coal was dug, but this makes coal more expensive than oil.

The same distributional problems occur with other energy sources. Each has noncontroversial applications and areas where it is cheap (solar power to heat swimming pools or hot water, wind power to pump water for cattle in isolated locations), but each is expensive and controversial when expanded to fill the enormous gap now filled with imported oil.

To think of alternative energy sources is to think of vigorous well-organized opponents. In some cases the opponents may be a minority, but they are perfectly capable of causing lengthy political and legal delays. The most visible are those who oppose nuclear power (cheap or expensive), but I have yet to meet anyone who wants a coal-fired, electrical-generating plant next to him. Environmentalists want coal mined and burned safely and cleanly without disturbing the topography at either end. It is not at all certain that this can be done, but it certainly cannot be done cheaply. Local communities in Montana and Wyoming don't want their life-styles disrupted to provide power for someone else. The small tests of modern, commercial, wind-powered generating plants have incurred neighborhood complaints about the noise from whining propellors. The only power sources that have not accumulated their own particular set of opponents are those that are so far from commercial application that no one yet fears their existence.

In each case the opponents fear a decline in their real standard of living. This fear is not an imaginary fear. They are right. If you are someone who places a high value on a clean environment, then a dirty environment is a substantial decline in your real standard of living. If you enjoy small-town western living, large coal mines are going to mean a major reduction in your standard of living. While everyone wants energy, everyone wants it produced in a way that does not disrupt their standard of living. Californians want their electrical power generated in Utah. Yet this cannot be done for all of us. If electrical power exists, it is going to disrupt someone's real standard of living. On average the gains will exceed the losses, but this average truth is of little recompense to those that must suffer the actual losses.

Each path also confronts the problem that energy independence is a time-consuming problem. No path or combination of paths could succeed in less than fifteen to twenty years. This gives us a classic political problem. The benefits of energy independence occur in the distant future, but costs, political and economic, start now. The politicians who must incur the costs and get reelected now won't be around to claim the credit when energy independence is achieved in the distant future. Why should they start down a path that is for them all pain? Our future economic life may be better if we start now, but we will have to pay now for those projects that will only later lead to energy independence.

But there is also another problem in the area of energy independence. The price of imported oil is high because of a cartel, not because of natural scarcities. Since the price is a man-made price, it becomes very difficult to make alternative energy investments. Suppose that imported oil is selling for $30 per barrel. Now imagine that you were able to discover some process that would make synthetic oil for $25 per barrel. Could you afford to go into production? If the $30 price were set by Mother Nature, the answer is clear "yes." A substantial profit could be earned over the full costs (including capital costs) of production. Given billions of barrels consumed, billions of dollars are available to be made.

If the $30 price is set by a cartel, however, going into production is not so clear-cut. Massive investments would be necessary to make the synthetic oil; but what would happen once you started producing oil? The cartel would simply cut their price and you would be left with a large worthless production facility. You would have done the world a favor–forced the price of oil down from $30 to $24.99 –but you would have lost your investment. Private investors usually do not do social favors that lead to capital losses. Since OPEC production costs (something like 40 cents per barrel in Saudi Arabia) are far below selling prices, a private investor cannot compete with OPEC. Private investors do not want to risk producing synthetic oil, and the price of imported oil stays at $30 per barrel.

As a result, energy independence is not something that can be achieved by liberating free enterprise. Because private companies are competing with a government cartel, they simply won't make the necessary investments. But this leaves us with a contentious choice. Investments in synthetic oil will only be made if they are made by a government corporation or with government subsidies to private corporations. Either technique requires taxing someone to raise the necessary revenue. Whose income gets cut? To go the public route is to meet the objections of those that are opposed to giving public money to rich private investors. Each side throws its political veto and we take neither route. But the result is energy dependence, with all of its instabilities and uncertainties.

In the face of gas lines in the summer of 1979 and political pressures to do something, Congress was drifting toward extensive subsidies for the production of synthetic liquid fuels. It remains to be seen, however, whether we will be able to raise the massive amount of revenue that will be necessary for this task. Will we really be willing to raise taxes by many billions of dollars to implement such a plan? Will we really be willing to give up on the environmental standards that make synthetic fuel very expensive? Will we really find communities where people are willing to live next to large, synthetic fuel plants? Maybe. But it is much easier to talk about building synthetic fuel plants and to pass legislation saying that they ought to be built than it is to do it. To do it you have to be willing to lower the real incomes of many people and many communities. No one wants to be those individuals or those communities.

We are in a morass from which we cannot escape. The president proposes timid energy legislation that could not possibly solve the basic problems, but Congress cannot digest even this. The basic problem exists, persists, and becomes more painful. But no solutions are possible since they all result in a shift in the distribution of income. Not having a clear idea of what constitutes a desirable distribution of income, we are unwilling to accept or ratify any of these changes. We have no way to decide *when* compensation should be forced to suffer real income declines. We sink because we will not swim.

Chapter 3

Inflation

IN ECONOMIC LIFE, unsolved problems have the unfortunate characteristic that they reinforce each other. In this case, our inability to solve the energy problem contributes to making the inflation problem much worse, since energy is used in the production and distribution of almost everything. Rising energy prices cause price increases in other goods and services. Wage earners attempt to keep up with prices by demanding large wage increases. This leads to even greater inflation. Whenever government policies seem to be moderating the rate of inflation, a new burst of energy inflation pushes the economy back onto the inflationary track.

The impact of energy prices on inflation is enormous. With 10 percent of our consumption going toward the direct and indirect purchase of energy, a 100 percent increase in the price of energy generates a 10 percent rate of inflation all by itself.[1] If we are to have a noninflationary economy in the presence of energy price hikes of this magnitude, the remaining 90 percent of our purchases must fall in price by an average of 11 percent. An 11 percent decline in 90 percent of our consumption just counterbalances a 100 percent price increase in 10 percent of our consumption. But an 11 percent price decline in 90 percent of our production means that everyone who is engaged in producing goods and services other than energy must suffer an 11 percent decline in his income.

This presents government policy makers with an economic problem and a political problem. Economically what policies can be imposed to cause 90 percent of our prices to decline by 11 percent?

Politically how can you impose economic policies that are going to reduce the vast majority of the population's income by 11 percent? Unless such policies can be imposed, the economy is going to have a significant rate of inflation as long as energy prices are rising.

Inflation is the paradigm zero-sum game. Whenever a price goes up, two things happen. Whoever buys that particular commodity finds that his real income goes down. But someone also gets that higher price, and his income goes up. That someone may be the seller, the producer (capital or labor), or the owner of raw materials, but no income disappears. For every loser there is a winner. Inflation can redistribute income, but it does not lower the total amount to be divided. Everyone cannot be worse off. Some individuals win; some individuals lose. This is not an economic hypothesis but algebraic necessity. Everyone wants a government that stops inflation, but one that does so by inflating his income and deflating the income of everyone else. To stop inflation in the presence of upward price shocks, such as energy, governments must adopt policies that lower someone's income. The problem is not finding economic policies that will lower incomes, but being able to impose them.

How Did We Get to Where We Are?

The history of our current inflationary predicament nicely illustrates the problems that emerge when government tries to avoid making distributional judgments. Like most severe problems, the history begins with two intertwining essential ingredients—bad luck and bad judgment. Suppose that you are in the process of climbing a mountain, are not roped together with your climbing partners, and an avalanche comes down the mountain. If you had been roped up (shown good judgment) you would have had a chance of surviving the avalanche. If you had not been roped up and there had been no avalanche (good luck), nothing would have happened and no penalty would have had to be paid for the bad judgment. In analyzing the causes of how our economy got to where it is, there are ample amounts of both bad luck and bad judgment. When this combination occurs, there is almost always a severe penalty to be paid. The penalty cannot be avoided. The problem should not be compounded by making the penalty larger than it has to be, but in this case, each attempt to avoid the necessary penalty made the problem worse.

The sequence of events that led to high inflation began with a bad judgment that now belongs to the long ago. To avoid making an unpopular war even more unpopular, President Johnson decided not to raise taxes when they should have been raised in 1965, 1966, and 1967. We were to have a war, but we were not going to pay for it overtly. Given a civilian economy that had just reached capacity operations before the Vietnam War began, the war created classic excess-demand inflation. The military, for example, needed millions of boots. Orders were placed, and this forced up the price of civilians's shoes as military boots competed for scarce leather and shoemaking capacity. Employers started competing for workers, and wages rose. All across the economy prices started to rise as demands exceeded production capacities.

This war was no exception to the rule that all wars must be paid for, but the payments came in the form of an implicit tax–inflation. Although the economic mistakes began in late 1965, it is important to note that inflation only gradually accelerated from 2.2 percent in 1965 to 4.5 percent in 1968.[2] Prices and wage pressures were gradually spreading across the economy, but it took a long time for these pressures to build up, even though the economic mistakes in financing the Vietnam War were very large. The economy does not respond quickly to either inflationary or deflationary pressures.

President Nixon's strategy for coping with the fruits of President Johnson's mistakes was to apply the classic economic medicine. Monetary and fiscal policies were tightened to induce a recession. According to economic theory, idle men and equipment would stop prices and wages from rising. Once prices and wages had stopped rising, the direction of monetary and fiscal policies would be reversed, and the economy would return to full employment by November 1972 (a magic date from the point of view of the president). The mild recession arrived on time in 1969 and 1970, but by the summer of 1971, the rate of inflation had not yet begun to fall. Instead it continued to accelerate to an annual rate of 5.9 percent in the first half of the year.[3] Unemployment stood at 6 percent.

There is every reason to believe that if President Nixon had continued his restrictive policies he could, in time, have stemmed the rate of inflation. But the inflationary momentum of the Vietnam War was so large that it was not going to be quickly stemmed unless the president was willing to incur an even bigger and much longer recession. But recessions hurt those who are unemployed. Their incomes fall and they quite naturally vote against those who are forcing them to be the economy's inflation fighters while others enjoy the fruits of their effort.

In the aftermath of even a mild recession, the public opinion polls showed that President Nixon would lose to his presumed Democratic challenger, Senator Muskie. Not wishing to run for reelection with high unemployment (which many blamed for his 1960 defeat) and inflation, the president dramatically changed his economic policies in August 1971. His only choice was to do so or court defeat.

Despite repeated promises never to impose price and wage controls, they were imposed to stop inflation while monetary and fiscal policies swung strongly toward stimulating the economy to lower the rate of unemployment. This double-barrelled approach was part of an extremely successful reelection campaign, but it merely postponed the basic problem. When controls were lifted in 1973, the inflation that had been suppressed by the controls in 1971 and 1972 reappeared and was intensified by the excess demand inflation produced by overstimulating the economy in 1972. We had been paying an economic price for President Johnson's decision to misfinance the Vietnam War, and we were about to start paying the economic price for President Nixon's reelection campaign.

But even this level of inflation was to be compounded by more bad luck and more poor judgments. Because of bad weather, crops failed in Russia. To raise the income of farmers and help in its reelection efforts, the Nixon administration sold too much wheat to the Russians in the summer of 1972. When Russian sales were subtracted from American supplies, there simply wasn't enough wheat to meet American demands, and prices rose sharply. Despite these shortages, the Agriculture Department left acreage controls in place for 1973. After twenty-five years of trying to dispose of surpluses, they simply could not believe that a period of shortages had arrived. When coupled with the 1973 corn blight, supplies fell even farther behind demands. Then the anchovies failed to appear off the coast of Peru, forcing European cattle feeders to shift from fish meal to American grains. The net result–a 66 percent rise in farm prices from 1971 to 1974.[4] Industrial inflation was now compounded with agricultural inflation.

Because other industrial economies were also growing rapidly in 1972 and 1973, raw material shortages were compounded by panic and speculative buying that led to even greater price increases than were warranted. The final blow was the OPEC price increase and the Arab oil boycott in late 1973. Imported oil prices tripled with corresponding price pressures on other energy sources. With energy prices an important part of the production and distribution costs of almost everything, significant cost pressures started to work their way through the economy. Prices went up because the cuts of imports went up. The economy now had both excess demand and cost-push inflation.

Given this sequence of events, the double-digit inflation of 1973 and 1974 is not surprising. Anything else would have been surprising. But the shock of double-digit inflation led to another sequence of events. Something had to be done about inflation. But what? The decisions of Presidents Johnson and Nixon could not be undone. Good weather and crops could not be legislated. United States economic policies could not reverse OPEC's tripling of oil prices. Not knowing what else to do, and being in political disarray with Watergate, the Nixon administration applied a very large dose of the classic medicine–tight monetary and fiscal policies.

The policies worked, in that real GNP stopped growing by the fourth quarter of 1973 and gradually fell throughout the first three quarters of 1974. Every quarter the GNP was getting smaller, larger and larger amounts of idle men and equipment were being created, but the rate of inflation did not respond. Since inflation did not respond, monetary policies were tightened further until they created the famous credit crunch of late 1974. Under the impact of tight money, demands fell rapidly. Home construction fell from 2.1 million units in 1973 to an annual rate of less than 900,000 million units in late 1974. A recession was created within a recession.

While the real GNP had fallen 2.5 percent from the fourth quarter of 1973 to the third quarter of 1974, it now started to plunge rapidly–falling another 3.2 percent in the fourth quarter of 1974 and the first quarter of 1975.[5] The end result was 9 percent unemployment. Instead of just inflation, we now had inflation and unemployment. The problem was now to recover from the sharpest recession since the Great Depression (but in an environment of rising prices rather than falling prices).

Partly because of the severity of the recession and partly because the adverse effects of any shock eventually wear off, the rate of inflation fell to 5.5 percent by mid-1975; but there it stuck despite massive amounts of idle capacity (30 percent in mid-1975) and idle labor (9 percent in mid-1975). After holding in the 5.5 to 6 percent raise for three years, inflation started to accelerate in 1978.

This accelerating is interesting because it illustrates the basic problem facing our government. Different groups–farmers, the elderly, the steel industry, low-wage workers–were demanding that government do something to give them more economic security. In response to these demands, government reintroduced a system of agricultural price supports, raised social security taxes to pay more benefits to the elderly, adopted reference pricing to protect the American steel industry from the Japanese, and sharply increased the minimum wage to help low-wage workers. The net result, (see table 3–1) of these and other similar actions, was to substantially raise the rate of inflation.

Other factors were at work–a rise in meat prices, the falling dollar–but our government caused more than half of all the extra inflation that occurred from 1977 to 1978. It did not cause this inflation because it was stupid, but because it was trying to raise the incomes of particular groups in our society. But to do so, government must raise prices, cause inflation, and reduce the incomes of other groups. With these and other groups making more demands, with rapidly rising oil prices, and with much of the 1978 inflation already built into the economy, the 1979 and 1980 prospects are for further

TABLE 3–1
What Made Inflation Worse?:
Sources of Accelerating Inflation in 1978

The Underlying Inflation Rate		
Per Annum: 1976–77		5.3%
Actual Change in Consumer Prices: 1978		7.7
Increase in Inflation Rate		2.4
Accelerating Effects: 1978		
Food Prices		0.7
Policy Measures	0.3	
Livestock	0.4	
The Dollar		0.4
Minimum Wage		0.1
Social Security and Other Policies		0.3
Home Ownership		0.6
Demand and Protection		0.3

SOURCE: Robert Gough and Robin Siegel, "Why Inflation Became Worse," *Data Resources Review* (Jan. 1979), p. 1.16.

acceleration in inflation. Some of this additional pressure comes from outside our economic system, but much of it is generated by demands for economic security within our economic system.

The Distributional Consequences of Inflation

Before looking at the different cures for inflation and who would be hurt if they were to be applied, let us look at the consequences of inflation itself. Whose income has gone up; whose income has gone down? What has happened to average incomes? In addition to telling us who has won and lost, an analysis of the distribution of income during a period of inflation also tells us something about the economy. The normal economic mechanisms for dampening inflation depend upon lags and shifts in the distribution of income. Some individuals find themselves with lower incomes and must cut their purchases. Demands fall, idle capacity emerges, and inflation halts. Sharp shifts in the distribution of income lead to a quick dampening of inflation; small shifts lead to a slow dampening. If relative incomes do not change, this is a major indication of an economy where inflation is not going to stop of its own accord. Demands do not fall, prices do not fall, and inflation continues as everyone raises his wages or prices at about the same rate.

Since inflation in its virulent form broke out in 1973, let's look at the performance of the economy pre- and post-1972. Starting from a 4 percent rate in 1972, inflation accelerated to 9.7 percent in 1974, fell back to 5.2 percent in 1976, and then reaccelerated to 8.5 percent in the first quarter of 1979.[6] Unemployment reached a cyclical low of 5 percent in 1973, rose to 8.5 percent in 1975, and then fell to 5.7 percent in the first quarter of 1979.

These gloomy statistics have been widely disseminated, as if they prove that the real standard of living has fallen. In fact, nothing of the kind has happened. From 1972 to 1978, real per capita disposable incomes rose 16 percent.[7] After accounting for inflation, taxes, and population growth, real incomes have gone up, not down. The average American is better off, not worse off. Nor is his real standard of living growing at a much slower rate than before. If you compare the real income gains in the six years since the onset of stagflation with the six years prior to stagflation, there is surprisingly little difference. In the go-go economic boom from 1966 to 1972, real per capita disposable incomes rose 17 percent. The real standard of living was rising slightly faster prior to 1972 than after 1972, but the difference is very small. Based on their own impressions, almost no one could tell the difference between a 16 percent and a 17 percent rise in living standards over a six-year period of time.

Why then are we in the middle of a period of national economic masochism where it is widely believed that the American standard of living is collapsing? One possibility is that the distribution of income is becoming unequal so rapidly, under the impact of inflation, that most Americans face a falling standard of living even though the total economic pie is growing larger. Averages are simply misleading. As we shall shortly see, this simply isn't the case. Distributional shifts are few and far between, and those that have occurred are very small. The averages are not misleading. Most Americans have experienced a real income gain of 15 percent from 1972 to 1978.

Why then do most Americans seem to be saying that they have been hurt by inflation? Part of the answer is due to money illusion. This is a disease that never afflicts a rational *homo economicus* but burdens almost every real human being. While real incomes were rising 16 percent, money incomes were rising 74 percent. Suppose a money man were to deliver $74 to your doorstep in the morning. You put on your bathrobe to go down to pick up the money along with the morning paper but find that when you get to your doorstep only $16 is there. Are you happy or mad? You are $16 better off than you were, but you have seen the $74 and can imagine what life would be like with it. You may even be able to convince yourself that your real standard of living has gone down. And in some psychological sense you may be worse off.

Money illusion is compounded by our puritanism. Everything we have we have earned. We have never been lucky. Whatever we have is due to our personal merit. As a consequence, when we see our money incomes rising we attribute it to our merit, neglecting to remember that inflation raises someone's income whenever prices rise. When we find that some of our money income gain is taken away, in the form of inflation, we see it as some alien hostile force taking away from us purchasing power that is rightfully ours. Each of us gives inflation credit for taking income away from us, but almost none of us gives inflation credit for raising our money incomes above what they otherwise would be. A world with a 74 percent gain in real incomes every six years would be a much less frustrating world than one with a 16 percent gain, but it also is an unattainable, imaginary world. If inflation had been halted, we would not have had $74 to spend. The money man would have only delivered $16. But this does not stop everyone from thinking that he is one of the losers. Government should do something to protect his income position.

Inflation also turns personal problems into what appear to be social problems. At any point in time in an economy as complex as ours, there are a wide variety of economic changes taking place. Many people are suffering real income losses and many people are making real income gains. In an economy without inflation, those who suffer reductions in their real income also suffer a reduction in the money incomes that they earn in the marketplace. They may blame this loss on bad luck, bad judgments on their part, or some other circumstance or economic factor, but they cannot blame the whole system, since the system is not allocating income reductions to everyone.

In a period of inflation, the same changes are occurring. Shifts in the basic supply and demand conditions of the economy cause the real incomes of different individuals to rise and fall. But inflation causes most people's money incomes to rise. Real incomes will still fall for some, since the rate of inflation exceeds the rate of increase in money incomes; but now each individual can think that if only inflation had not occurred, he would have had a rising real standard of living. After all, his money income has risen. As a result, what is, in fact, a personal problem is seen as a social problem. Many college professors blame inflation for their falling relative incomes, while the real cause lies in demography and surplus Ph.D.'s.

Some misleading statistics also contribute to the illusion. We are all familiar with the idea that the real take-home factory wage has gone down. Headlines regularly proclaim it to be so. What the headlines do not tell us, however, is that less than 20 percent of the American work force is a factory worker, and that the averages are down not because of inflation but because of the rise of part-time workers. Because of rising female participation rates and part-time work, average hours of work are going down, and this leads to lower average earnings even if hourly earnings are going up.

From 1972 to 1978 our real GNP grew by $228 billion (1972 prices).[8] If a larger economic pie exists, as it does, then someone is getting that larger economic pie. With $228 billion more to go around, someone has to have a much higher standard of living. Unless there have been shifts in the distribution of income, most of us are participating in that larger economic pie.

Table 3–2 shows the distribution of money income for households from 1972 to 1977 (the last year for which dates are available).[9] Given the size of the measurement errors with this kind of data, there has been essentially no change in the distribution of money income between rich and poor. The top 40 percent of the population had 69.5 percent of total income in 1972 and 69.6 percent in 1977. The bottom 40 percent of the population had 13.7 percent in 1972 and 13.8 percent in 1977. While the changes are not large enough to be statistically significant, there may have been a very small shift in income toward the bottom 20 percent and the fourth quintile (sixtieth to eightieth percentiles).

TABLE 3–2
Distribution of Money Income Among Households (%)

Quintile	1972	1977
First	4.1	4.3
Second	10.5	10.3
Third	17.2	16.9
Fourth	24.5	24.7
Fifth	43.7	43.8

SOURCE: U.S. Bureau of the Census, *Current Population Reports, Consumer Income 1977,* Series P–60, no. 117 (Dec. 1978), p. 19.

While the distribution of money income has not been altered by inflation, the charge is often made that the distribution of real income has been altered since the cost of living has gone up faster for low-income groups which spend more of their income on those goods and services (food, fuel, and so forth) that have gone up the most in price. When cost-of-living indexes are calculated for each of the five quintiles, this charge is not substantiated. From 1972 to 1977 the implicit price deflator for personal consumption expenditures rose by 39 percent. No quintile, however, has experienced a rise in cost of living that is more than five percentage points above or below this average. Converting the money distribution of income to a real distribution of income does not change the conclusion that there has been essentially no change in the distribution of income between the rich and the poor. With both money incomes and the cost of living being essentially constant across the distribution of income, simple algebra leads to the conclusion that all income groups have experienced a 15 percent rise in their real standard of living.

Another argument that is often heard revolves around the progressivity of the federal income tax. In a period of inflation, individual taxpayers move up the progressive rate structure since taxes are levied on money income. This leads to a higher level of taxes on real income. While the argument is certainly correct if everything else remains the same, everything else does not remain the same. More individuals itemize their deductions (which also rise with inflation) and Congress periodically cuts taxes to offset the impact of inflation.

When one actually analyzes the data, the latter effects more than offset the former effects. Federal personal income taxes were 10.9 percent of personal income in 1972, but they only averaged 10.5 percent from 1972 through 1978.[10] Since the onset of inflation, the federal income tax burden has fallen, not risen. While total tax collections are not up, the distribution of tax collections could have shifted to increase the burdens on some income classes. Here again, there is no indication of any change. In the most recent data, each quintile of taxpayers is paying the same share of total taxes that they paid in 1972.

The same absence of any shift in the distribution of income is seen if one looks across all of the normal socioeconomic groups. From 1972 to 1977 the percentage of the population living below the official poverty line (corrected for inflation) has declined from 11.9 percent to 11.6 percent.[11] Black household incomes have risen from 58 to 59 percent in comparison to the income of whites. Data on Hispanic households are not available for 1972, but their relative income has risen from 71 percent to 75 percent in comparison to whites's relative income from 1973 to 1977. Low-income groups have not gained in the 1970s as they did in the 1960s, but they also have not been falling behind. High unemployment and slower real growth have stopped them from catching up with the mainstream of the economy, but these factors have not forced them back to the levels of relative income that existed in the early 1960s.

Since elderly incomes are much more skewed than the incomes of the rest of the population, the relative position of elderly households depends upon whether you look at mean or median household incomes. Per capita mean household incomes for the elderly fell insignificantly from 94 percent to 93 percent of the entire population's from 1972 to 1977.[12] Per capita median household incomes rose slightly from 72 percent to 75 percent of the entire population's over the same time period. Social Security payments are also heavily underreported in our official income data. If one corrects for this underreporting, in 1972 and 1977 per capita mean elderly household incomes hold constant at 100 percent of the rest of the entire population's. Per capita median household incomes rose from 80 to 85 percent of the entire population's. If anything, the elderly have improved their relative income position during the period of high inflation.

The same conclusions hold if one looks at other groups such as farmers and the young or at the split of income between individuals, governments, and corporations. What changes have occurred are all very small and would probably have occurred in any case. This is not surprising in an economy where government raises income transfer payments to protect the real income of the poor and the elderly, and where labor unions index their wages to keep pace with inflation (see below). Everyone keeps up with inflation because everyone is part of the cause of inflation. The sharp shifts in the distribution of income that are often condemned in discussing inflation simply do not exist when one looks at the fact. This does not stop inflation from being a serious problem, but it does mean that many of the policies often suggested to stop inflation–a sharp recession–can have more adverse distributional consequences than the disease itself.

There is, however, one exception to the dictum that there has been little or no change in the distribution of economic resources in the 1970s. The distribution of net worth (wealth) has become more equal because the stock market has fallen. Since the richer you are the larger corporate stock looms as a fraction of your portfolio, the richer you are the greater your losses are on corporate stock. And in real terms, the value of corporate stock has fallen over 50 percent since 1968. But the decline started well before the onset of rapid inflation and has been slower since rapid inflation began in 1973.

Analysis of declining real stock prices by two MIT economists indicates that they are probably related to mistakes in how the financial community evaluates stocks during a period of inflation rather than to the effects of inflation on income flows.[13] Investors forget to correct for the fact that the real value of corporate debt declines during a period of inflation. And they discount earnings, with interest rates that include the impact of inflation without a corresponding increase in future expected earnings.

If you look at the distribution of the GNP, it is clear that the period of inflation that we have had since 1972 has made surprisingly little difference. This tells us two things about the economy. First, the impact of high unemployment has been extensively cushioned by government transfer payments. Second, inflation seems to have had little, if any, impact on the distribution of income. This means that all wages and prices are rising by about the same amount. Inflation has helped and hurt each of us. There have been no dramatic shifts in the distribution of economic resources since the onset of inflation.

None of this analysis should be taken to mean that inflation can be ignored as a source of real economic pain. While groups have not been hurt, individuals undoubtedly have been hurt. To say that the system can cope with inflation is not to say that inflation is desirable. We might have performed even better if inflation had not existed. At the same time, the lack of dramatic distributional shifts and the existence of a good real rise in living standards for most Americans should tell us something about choosing among the different cures for inflation. Since all of the potential cures involve substantial costs, it is important to choose a cure whose costs are less than those of the disease itself.

The consistency in the distribution of income also indicates that the economy is not going to quickly dampen inflation if it is just left alone. Real demands will not fall and stop prices from rising since real incomes are not falling for any significant group.

The Structure of a Modern Economy

Inflation is endemic in a modern economy for a very simple reason. Whenever upward price shocks occur, inflation will occur unless other prices and incomes fall. But in a modern industrial economy, prices and wages in other sectors do not easily fall. There is a substantial amount of downward price rigidity. This is due to both the structure of the private economy and the actions of government. Falling prices mean falling incomes, each of us organizes publicly and privately to ensure that we are not the ones who will experience falling prices and incomes. But if each of us is successful, inflation will be the inevitable result.

For a number of reasons, it takes great shocks to reduce industrial wages and prices. Government is one of them. If the Japanese are threatening to force down the prices of steel, textiles, and electronic goods, our producers run to the government for protection to prevent their prices and incomes from falling. Sectors that experience fluctuating free market prices (agriculture) demand and get government interventions that stop prices from falling. Minimum wage rates and prevailing rates on government construction contracts place floors under the wages for workers.

Large companies and unions play a role. Both have the power to stop their prices or wages from falling. Union leaders do not have to worry about the votes of the unemployed. The employed want higher wages. Oligopolistic firms are aware that they have more profits to lose from cutting prices than from cutting output. And as long as the demanded cutbacks in production are modest and of short duration, they can informally coordinate such cutbacks with their industrial competitors. But corporate and union power is not absolute. Prices and wages do not respond to moderate amounts of excess capacity, but they would fall if demands fell enough. To re-create the Great Depression would be to once again see falling prices.

In our economy wages are even less flexible downward than industrial prices. Unions play a role in this, but with only 20 percent of the labor force unionized, unions cannot be blamed for wages that do not fall in the remaining 80 percent of the economy. The basic problem is that stable wages are useful in a modern industrial economy. It is irrational for private employers to take advantage of unemployment and lower the wages of their existing work force unless the downward shift in demand is so large that they are forced to do so to avoid bankruptcy.

Standard free market economics is based upon four basic assumptions about the characteristics of the labor market: (1) Skills are exogenously acquired and then sold in a competitive auction market; (2) The productivity of each individual worker is known and fixed; (3) Each individual worker's happiness with his wages depends solely upon his own wages. Workers never look to see what others are getting; (4) Total output is simply the summation of individual productivities. If the economy's actual operations were based on these four hypotheses, wage inflation and unemployment could not exist. Whenever unemployment arose—an unexpected decline in demand or a faster than expected rise in the labor force—wages would fall until the extra workers were reabsorbed into employment or until they had decided that wages were no longer high enough to merit sacrificing their leisure. But wages do not respond to moderate amounts of unemployment, and unemployment is extremely persistent.

Upon examination, the basic assumptions about the nature of the labor market seem less than adequate. They ignore long-run employer-employee interests in a good mutual relationship.

1. They ignore the fact that much of our human capital is acquired on the job rather than in formal education. This can be seen in the analysis of the determinants of earnings or in the surveys of where working skills are acquired.[14] The labor market is not primarily a market for allocating skills but a market for allocating training slots. Workers are only trained when job openings exist and an independent supply curve does not exist. But without independent supply and demand curves, wages must be determined in some fashion other than by a market correction.

2. Instead of being fixed and known, individual productivity is variable and difficult to know. Each worker has a maximum productivity, but depending on motivation, he can provide any productivity between that maximum and zero. Employers find it difficult and expensive to know how much productivity each of their employees is providing. And even if they do know, it is difficult and expensive to change wage rates or fire an employee. As a result, every industrial operation requires a substantial component of voluntary cooperation. If employees choose to withhold that voluntary cooperation (work to rule), any industrial operation in the country could be brought to its knees. Evidence for the potential

variability in productivity can be seen in the discipline of industrial psychology, business interests in motivation, and through introspection.

3. Instead of having independent preferences, most workers have interdependent preferences where their satisfaction depends on their income relative to that of their neighbor's. Evidence for this can be seen in the sociological literature on relative deprivation and economic surveys of what causes economic satisfaction.[15] These surveys universally find that people's satisfaction or dissatisfaction with the economic circumstances is dependent on their relative income and not on their absolute income.

4. All industrial operations are subject to a substantial component of team as well as individual productivity. Evidence for this can be seen in the sharp learning curves of new industrial plants. As workers learn to work with each other, costs of production fall dramatically. They develop teamwork and team productivity that is over and above their individual skills and individual productivity.

But under these circumstances, where does economic analysis lead? It leads to the two factors that are widely observed in the labor market: (1) Money wages exhibit downward rigidity; they do not fall when surplus labor exists; (2) Relative wages are rigid and change only in the long run.

Because skills are acquired on the job, in an informal process of one worker training another, every industrial operation needs workers willing to be trainers. But in a truly competitive world, no one wants to be an informal trainer. Every worker realizes that every additional worker trained will result in lower wages and a greater probability of being fired in any economic downturn. It is rational in a competitive world for each individual to seek a monopoly on local knowledge (how to run machine *X*) and then refuse to share his or her knowledge with anyone else. This preserves wage and job opportunities. To promote training and make workers willing to be trainers of other workers, employers essentially offer two guarantees. First, they promise not to lower wages if surplus workers become available. Second, they promise to hire and fire based on seniority. This means that each trainer's trainees will be fired before he is. Essentially the employer agrees not to be a short-run cost minimizer in the interests of long-run training and efficiency. But this leads to money wages that do not fall when unemployment emerges.

Rigid relative wages spring from interdependent preferences, but these preferences are enforced on the employer through the employee's ability to vary his own productivity and disrupt team productivity. Because of interdependent preferences, workers perceive some wage differentials as fair and other wage differentials as unfair. But they need some threat to force employers to pay fair, relative wages. In the textbook economy, there would be no way to enforce the interdependent preferences even if they existed. But in the real world, employees can cut their own productivity or disrupt team productivity if wage differentials are perceived as unfair. Employers find it difficult and expensive to determine who is providing less productivity. Even knowing who is at fault does not lead to an easy solution. Firing is expensive and disrupts the team. As a result, employers find it more profitable to pay the wage differentials that employees view as fair than to shift to the wage differentials called for by changes in short-run supplies and demands in the labor market. Total productivity paying "fair" differentials is higher than total productivity paying supply and demand differentials since workers can alter the level of productivity depending on their satisfaction or dissatisfaction with pay scales. The net result is a structure of rigid relative wages that do not fall when unemployment emerges.

This pattern exists in any industrial operation, but it has been highly visible on the nation's sporting pages over the past few years. One superstar gets a large wage increase, and this leads other superstars to break their existing contracts for higher wages. With superstar wages rising, lesser players demand and get large wage increases. Each threatens to use his power to disrupt the team and lower his own productivity if his demands are not met. Wages rise and owners pass the burdens onto consumers in the form of higher ticket prices. While the wage structure and bargaining of sporting teams is more visible than most, it is by no means unique. What goes on there goes on in a milder form throughout our complex interrelated economy.

Wages are set in a social process that is far removed from simple supply and demand curves in a modern industrial economy. From the employer's perspective this process is inefficient in that he cannot adjust wages to individual productivities and short-run changes in circumstances, but it is efficient since his production team is not disrupted by dissatisfied workers, and since training occurs at less cost than it would otherwise. The gains from rigid wages are greater than the gains from flexible wages.

With downward rigidity in money wages and fixed relative wages, labor markets cannot clear via wage reductions and shifts in relative wages. They clear based on worker qualifications (level of education and so forth), but this leaves the economy with unemployment and inflation. Workers who are willing to work at current wages cannot find work. Since wages do not fall, prices do not fall. Instead of reducing prices in times of excess capacity, firms cut production. This produces more unemployment and exacerbates inflation. When oil prices rise, other prices do not fall. If anything, other prices and wages rise in an effort to catch up.

With rigid wages, the demand for economic security becomes more comprehensible. In a world of wage flexibility everyone has economic security in that one can always find a job by being willing to work for slightly less. In a world of downward wage rigidity where skills are acquired on the job, economic life is much more uncertain. You may end up unemployed, and you do not control the skills you will learn. To a great extent they will depend upon what learning opportunities are allocated to you in the job market.

But there is another major factor that leads to increasing inflation. It is the phenomenon of *indexing*. Since 1974 and the scare of double-digit inflation, labor, business, and government have sought to protect themselves from the uncertainties of future inflation by adding cost-of-living indexes to all of their future commitments. Cost-of-living escalators are increasingly being built into government wages and programs. Very few business contracts are currently signed without the protection of inflation escalators. Cost-of-living clauses have become almost universal in new labor union contracts since 1974. Nonunion workers do not have legal cost-of-living clauses, but companies that provide cost-of-living protection to their unionized workers almost always give the same protection to their nonunion employees. Similarly, nonunion employers de facto index wages to keep their best employees from moving to employers who do index and to keep unions out. Adding cost-of-living escalators to private contracts is a perfectly rational response to inflation on the part of both business and labor, but it fundamentally alters the nature of the economy and the effectiveness of monetary and fiscal policies.

The classic objection to "legal" indexing is that it reduces the effectiveness of monetary and fiscal policies. The reasons are easy to understand. If inflation is 6 percent this year, all wages and prices will go up 6 percent next year due to escalator clauses; but this leads to a 6 percent rate of inflation next year and hence to 6 percent increases in wages and prices in the third year. While only parts of the economy (some government programs and wages) have been legally indexed, there is no difference between government indexing and private indexing when it comes to their impact on macroeconomic policies. The policies still produce unemployment, but they lose their capacity to reduce inflation. They can only reduce the rate of inflation if they are tight enough to produce a basic wage settlement (excluding the cost-of-living clauses) less than the rate of growth of productivity (about 2 percent). Given basic settlements currently far in excess of that level, unemployment would have to be much higher to generate the appropriate settlements.

Indexing is one of the main reasons why the inflation rate stuck at 5.5 percent per year in mid-1975 and did not fall further, despite large amounts of idle capacity. Not knowing what was going to happen in the future, economic actors did the rational thing. They inserted an insurance clause (a cost-of-living escalator) in their agreements about wages and industrial prices. But in the process, they changed the characteristics of the economy. Whatever the degree of price and wage responsiveness to idle capacity before indexing, it was less after indexing.

But there is an added problem. Any upward price shocks will be built into the index and carried forward into the future. If further oil price increases lead the rate of inflation to rise from 8 percent to 12 percent, then indexing will carry the 12 percent inflation forward into the future since wages and prices will now rise by 12 percent per year rather than by 8 percent per year.

With prices and wages inflexible, downward, and indexing, the monetary authorities lose most of their power to stop inflation.

To stop inflation in the face of upward price shocks someone's prices and wages must go down, but tight money no longer leads to this result. The monetary authorities are confronted with the choice between letting the money supply grow and confirming the rate of inflation or stopping the money supply from growing and producing unemployment to go along with the inflation. But they do not have the power to stop inflation—to make other prices and wages go down without producing a major recession or depression.

This structure of our economy emanates from a simple human desire. Although falling wages and prices might be good for the economy, they are not good for the individuals or groups whose income falls along with these falling wages and prices. Each of us organizes to avoid being subject to falling prices. But if we all succeed, we have an economy where inflation is endemic. To stop inflation someone's income must go down.

Potential Cures

Given a heavily indexed economy where prices and wages do not fall subject to occasional upward price shocks, what are the options? There are essentially five, but each of them has severe drawbacks. Those that could potentially cure the problem have the basic characteristic that they would substantially lower the real income of some significant group in our society.

Since no overt choices have to be made, the easiest option is to simply tolerate the current inflation. And until we learn how to choose some option overtly, this is exactly what we will do. Countries like Brazil and Japan have demonstrated that it is possible to grow rapidly with high rates of inflation. What small, adverse distributional effects do exist could be made even smaller by indexing those parts of the economy that are not indexed. Unemployment is high, but it is concentrated among the young. If we are patient, many believe demography will sharply reduce the number of young people and the unemployment problem may cure itself.

Against this one can argue that we have to be concerned about income losses to isolated individuals even if groups don't lose, and that the current system disrupts long-run economic planning. While the number of young people will decline, those who are now young will continue to exist, and high unemployment may plague them throughout their working lives. As they grow older the consequences of unemployment obviously become more severe.

The most telling objection to this option, however, is the simple fact that tolerating inflation is untenable politically. Americans want something done. Those with falling real incomes think that their problems are due to inflation. Those with rising real incomes think that their incomes are falling or that they would rise even faster without inflation. Being voters they want their elected representatives to do something. Doing nothing may not make real economic insecurity worse, but it seems to make economic insecurity worse. Sooner or later the demand to do something will express itself. What we do may not cure the problem, but we will do something.

The first real cure would be to tighten fiscal and/or monetary policies to the point that they created a recession or depression large enough to crack indexing, stop inflationary expectations, and force wages and prices to fall. While we could argue about exactly how high unemployment would have to be for these effects to occur, no one doubts that there is some level of unemployment that would stop inflation.

West Germany and Switzerland are often held up as countries that followed this route and succeeded in stopping inflation. They did so, but each country had two advantages that we do not have. Because the fixed exchange rate system of the post–World War II period had led to their currencies being undervalued, the currencies were appreciating in value in the aftermath of the double-digit inflation of 1974. With rising currency values, the price of imports was falling and served to moderate inflationary shocks. As we have seen, the Swiss ended up with cheaper oil after 1974 since the value of the Swiss franc went up faster than the dollar price of oil. Appreciating currencies are nice to have in an inflationary period, but one country's appreciation is another country's depreciation. Those with depreciating currencies find that import prices go up even faster. More importantly, West Germany and Switzerland each run an economy with a large number of foreign workers. When tight monetary policies lead to falling employment, the unemployed can be exported to the countries from which they came. And this is exactly what was done in West Germany and Switzerland. In 1978 industrial employment was 12 percent below 1973 levels in West Germany and 10 percent lower in Switzerland.[16] This did not lead to massive unemployment since each country sent foreign workers home. Scaled up to an economy the size of ours, the Swiss rounded up 10 million workers and sent them home. Which 10 million American workers do we round up and send home?

There is no doubt that inflation was temporarily cured, but the price was enormous in terms of unemployment. Given our rising labor force and unemployed workers who cannot be sent home, similar policies in the United States would have produced an unemployment rate approaching 30 percent. I have no doubt that prices would fall. But we would also be in the midst of another Great Depression with all its enormous economic losses and intense political pressures. Nor does the "big bag" solution ensure you against future inflation. With the rise in oil prices in the spring of 1979, U.S. inflation rates rose by three percentage points, but inflation rose by almost five percentage points in West Germany.[17] Each inflationary shock needs ever-tightening monetary and fiscal policies. At some point even West Germany runs out of foreign workers to send home. At this point tough choices need to be made. Disemploying West Germans is a different matter from sending foreign workers home.

A severe recession and high unemployment will cure inflation, but the costs are very unevenly carried by those groups that actually suffer from unemployment. This structure can be seen in the actual structure of unemployment for 1978. Officially unemployment ranged (see table 3–3) from 8.4 percent for teenage black females to 2.6 percent for white males aged fifty-five to sixty-five; [18] but if one corrects for the number of black teenagers who have dropped out of the system (are not at school, at work, or looking for work), black teenage rates rise to 52.7 percent and approach 90 percent in some central city areas. Almost 50 percent of those officially unemployed are from sixteen to twenty-four years of age.

TABLE 3–3
Structure of Unemployment in 1978

Males	5.2%
White Males	4.5
16–19	13.5
20–24	7.6
25–54	3.0
55–64	2.6
65 & up	3.9
Black Males	10.9
16–19	34.4
20–24	20.0
25–54	6.6
55–64	4.4
65 & up	7.1
Females	7.2%
White Females	6.2
16–19	14.4
20–24	8.3
25–54	4.9
55–64	3.0
65 & up	3.7
Black Females	13.1
16–19	38.4
20–24	21.3
25–54	8.7
55–64	5.1
65 & up	5.0
Hispanics	9.1
16–19	20.6
Males 20+	6.3
Females 20+	9.8

SOURCE: U.S. Department of Labor, *Employment and Earnings* 26, no. 1 (Jan. 1979): 160.

At the beginning of 1979 many analysts were arguing that the economy was at full employment despite a 6 percent rate. Even the 1979 *Economic Report of the President* expressed some sympathy with the position that we were approaching full employment for prime-age (twenty-five to fifty-five) white males. While one can quibble as to whether a 3 percent rate for prime-age white males is really full employment for this group (their unemployment rate reached 1.2 percent in 1969), there is no denying a very uneven structure of unemployment. Women, adult blacks, Hispanics, elderly whites, and young whites range between the extremes of black teenagers and prime-age white males. And as the national rate rises this dispersion becomes even larger.

Sometimes unemployment is simply dismissed on the grounds that it consists of millions of individuals unemployed for very short periods of time that are of no consequence to any of the individuals involved. This simply does not square with the facts. Over 50 percent of the total number of weeks of unemployment is borne by individuals who are unemployed more than twenty-seven weeks. Almost 50 percent of all spells of unemployment end up not in employment but in withdrawal from the labor force. A discouraged worker who has actively quit seeking work is not a mark of economic success.

If labor were sold in a competitive environment with flexible wages, such a structure of unemployment could not exist. Labor shortages would cause the relative wages of prime-age white males to rise and labor surpluses would cause the relative wages of other groups to fall. Employers would respond to these changes in wages by hiring fewer prime-age white males and more of the rest of the labor force. Unemployment gaps would shrink. In fact, the uneven structure has not been disappearing but has been growing worse during the 1970s.

If a recession is to be used to stop inflation, we draft inflation fighters in a very uneven pattern. In proportion to their size in the labor force, sixteen to twenty-four-year-olds are three times as likely to be drafted as adults. Females are 38 percent more likely to be drafted than males; blacks are twice as likely to be drafted as whites; and Hispanics are 75 percent more likely to be drafted than whites.[19] The group that is drafted least is prime-age white males, but this is precisely the group where unemployment is most effective in causing wages to fall.

Often the argument is made that women and children should be more than proportionally drafted into the war on inflation because there are fewer harsh economic consequences when they suffer unemployment since they have families to fall back upon. I have yet to hear this argument made by women or young workers; it ignores a very large number of women and young workers who are family heads or come from low-income families. It also ignores the long-run consequences of having a generation of young people who have either dropped out of the economy or have not gained the work experiences that lead to skills in later life. What do we do when today's unemployed twenty-year-old becomes the next generation's unemployed forty-year-old?

A major recession would stop inflation, but it would exacerbate many of our fundamental problems. Economic security and economic growth would both have to be foregone. Declining incomes would be a reality. Individual virtues would no longer be overwhelmed by the social disease of inflation, but they would now be overwhelmed by the social disease of unemployment. The distribution of income would shift sharply toward inequality as the economically strong forced the economically weak out of the economy.

The second real option is to impose formal wage and price controls—a collective defense against inflation. What the controls have to do is clear. Actually doing it is the problem.

Imagine an indexed economy with an 8 percent rate of inflation. If everyone agreed to raise his or her wage or price by only 5 percent, instead of the specified 8 percent, no one would be worse off and the inflation rate would be reduced to 5 percent. How to accomplish this, however, is not clear. Every individual economic actor has an incentive to raise his wage or price by the full 8 percent, since he will have increased his real income by 3 percent if the rest of the world goes down to a 5 percent gain. Conversely, if he cooperates with the incomes policy and goes down to a 5 percent gain while everyone else stays at 8 percent, he will have made a 3 percent income loss. Thus there is no such thing as a voluntary incomes policy. The incentives not to cooperate are simply too large.

The problem is similar to that found at a football game. Suppose an exciting play takes place. To get a better view individual spectators stand up; but if everyone stands up no one gets a better view and now everyone is uncomfortable since they have to stand rather than sit. But the first one to stand up gets a better view until everyone else stands up. Only collective action can keep everyone seated; individual decisions will lead to everyone standing. But what about the process of sitting down? The first person to sit down gets the worst view, and the last person to sit down gets the best view. Everyone wants to be last and everyone stands.

The same is true with inflation. The first person to raise his prices and the last person to stop raising his prices are the winners in the inflation game. Everyone wants to try to be that individual who will win and everyone wants to avoid being that individual who will lose. Farmers at one and the same time can be screaming about inflation yet demand agricultural price supports to raise farm prices. Everyone wants inflation in his own prices and wages and deflation in everyone else's prices and wages.

But there are problems beyond those of noncooperation. Whenever the controls are imposed, there will be some groups that are ahead of other groups. An even phase-down of inflation will leave those that started the inflationary process permanently ahead. There will be other areas where basic supplies and demands call for raising prices. Exemptions from the controls will be needed and will be used. But every exemption leads those who are not exempt to wonder why they should be the economy's inflationary fighters. If some of the inflation is due to the price of imported oil, controls must lower the incomes of some Americans to pay foreign energy producers. Whose incomes should controls lower? No one wants it to be their income.

Controls are opposed because groups believe that they will lose if controls are imposed. The business community is fond of attacking general price and wage controls on the grounds that it is "un-American," but they are the first group to ask for controls when it can help them raise their prices. They simply fear that general controls will be used to squeeze their incomes and not someone else's incomes.

The currently fashionable form of incomes policy discussion revolves around "tax-based incomes policy." Employers would be given a series of tax incentives or penalties, depending on whether they did or did not live up to some enunciated standard of noninflationary behavior. A tax-based incomes policy is, however, just equivalent to a set of wage and price controls with a predetermined set of financial penalties for violators. Catching the violators and enforcing the rules is no less difficult or expensive. The system is more flexible (if you want to violate the rules you can pay your penalty and violate the rules), but it is every bit as complex and expensive to administer. Detailed norms must be written and then enforced. In the Korean War, eighteen thousand price and wage inspectors were necessary to make the system work. There have been advances in computational techniques, but the economy is now much larger than it was then. Any serious system would require a large number of employees. There is no such thing as wage and price controls without a large bureaucracy to administer them.

While controls can be made to work, any compulsory incomes policy is more difficult in a democratic peacetime economy. Even with eighteen thousand price inspectors during the Korean War, the system needed widespread voluntary cooperation. Without some external threat, it is difficult to envision the necessary degree of voluntary cooperation.

The program is also technically more difficult. In wartime it is relatively easy to see how prices should be adjusted. They should be adjusted to maximize military production and minimize civilian production. But in peacetime the goals are not that simple. Which of all of the millions of civilian commodities should be expanded in supplies, and which should be contracted? Wage and price controls in wartime are also accompanied by labor, investment, and production controls. These controls make it easier to enforce wage and price controls because those who evade one set of controls will probably be caught in another set of controls. In peacetime wage and price controls are adopted without investment, production, or labor controls.

The real objection to controls is not that they are cumbersome and inconvenient (they are) or that they won't work (they will), but that they must reduce someone's real income if they are to succeed in stopping inflation. It isn't possible to predict who this will be without knowing the exact details of any system of controls, but there is no doubt that someone will be hurt. Some groups are vigorous opponents of controls because they believe that they will be the ones to suffer reduced incomes. Those groups that think they would gain under controls want controls.

In the third potential cure, government attempts to balance upward price shocks with downward price shocks. If the price of energy rises, the government looks around the economy to see where it has leverage to reduce prices. If inflation is already underway, government attempts to jolt the economy with a series of negative price shocks that will become embedded in the structure of indexing and spiral downward to a permanently lowered rate of inflation.

The deregulation of the airlines industry (and the resultant reduction in air fares) is one such program. If deregulation can lower air transportation costs, this lowers the measured rate of inflation. With a lower inflation rate and indexing, all wages and prices will go up less than they otherwise would have, and the future rate of inflation is less than it otherwise would have been. Deregulation could force prices down in trucking and many other industries. Instead of raising social security taxes (a tax which shows up in the cost-of-living index) and cutting income taxes (a tax that does not show up in the cost of living), all tax cuts could be focused on those taxes that reduce the cost of living, and all tax increases could be focused on those taxes that do not show up in the cost of living. The postal service could be made subject to private competition and forced to cut prices.

There are a host of government programs designed to raise prices that could be abandoned. Such programs now exist in agriculture, the maritime industry, the steel industry, textiles, and many others. Abandoning any or all of these programs would substantially reduce the rate of inflation. Efforts can also be made to restructure bottleneck industries where inflationary pressures are endemic, such as health care. Health insurance systems can be structured to encourage cost savings and discourage overusage and administered price increases.

Given the current upward momentum in inflation and the expected future upward shocks from energy prices, government policies would have to create *massive downward price shocks.* Massive downward price shocks is simply an economist's way of saying that many economic groups are going to be forced to accept sharp declines in their real incomes. Whenever government forces prices down, it forces someone's income to go down.

While the economics of the third anti-inflationary strategy are clear, the politics are not. Each of these anti-inflationary actions has a vigorous set of opponents. Truckers do not want trucking deregulated. Upper-income groups would rather have income tax reductions than sales tax reductions. State and local governments want their grants-in-aid without strings. Each of the industries protected by government policies to raise prices wants these policies kept in place. No one wants to volunteer as a price fighter in the war on inflation.

While everyone is in favor of reducing the rate of inflation as long as this is accomplished by lowering someone else's income, everyone is also against any anti-inflationary policy that lowers his or her income. Unfortunately an effective anti-inflationary policy has to lower someone's income below what it otherwise would be. This is not a matter of economic analysis but simply an algebraic truism.

Whose income is going to go down given that the income of energy producers is going to go up?

As a result, negative price shocks represent a real threat to the economic security of many individuals and firms. They may stop inflation and they may be good for economic growth, but they represent a real threat to the economic security of those whose prices and incomes are going to be negatively shocked.

Lacking a consensus of whom to hurt in the fight on inflation, the Carter administration has pursued an anti-inflationary strategy which is a moderate amalgam of all three of the basic strategies. The prime ingredient in the Carter strategy is monetary and fiscal policies designed to create a substantial, but still moderate, amount of idle capacity. Whenever the idle capacity threatens to disappear, policies are tightened to slow the economy and re-create idle capacity. Supposedly this idle capacity will gradually slow the inflation rate. We are told that it took a long time to build the current inflationary pressures, and that therefore we must expect it to take a long time to eliminate those inflationary pressures.

The prime strategy of idle capacity is to be augmented with voluntary wage and price controls that are supposed to reduce the rate of inflation slightly faster than would have happened if we relied solely on idle capacity. Hence the price controls are set to cut one half of 1 percent off the rate of price increase of the previous year, and the wage controls are set to lower the rate of wage increase slightly below that of the previous year.

At least a verbal attachment to the third strategy is also shown. We are reminded that airline deregulation went into effect under the Carter administration (the process began much earlier) and we were told that further deregulation (trains and trucks) would be proposed. Here again the strategy is to use moderate deregulation to shave the rate of inflation slightly faster than it might otherwise fall.

Patience may be a virtue, but it is a virtue which the American public probably does not possess when it comes to inflation. But even more importantly, a gradual phase-down is not possible in a heavily indexed economy with the mix of the three strategies chosen by the Carter administration. For the strategy to work you must believe that moderate amounts of idle capacity will gradually dampen price increases, and that perhaps this process can be slightly accelerated with voluntary wage-price controls and maybe a little bit of deregulation.

In a heavily indexed economy moderate amounts of idle capacity will not reduce the rate of inflation. Whatever last year's or last quarter's rate of inflation, it is carried forward into this year or this quarter by escalator clauses in wage contracts and in industrial contracts among firms. In an indexed world only three things can bring down the rate of inflation. If there are exogenous downward price shocks (due to a rising dollar or a falling oil price) the rate of inflation will gradually subside. If the basic wage settlement (leaving out the cost-of-living escalator) is less than the rate of growth of productivity, then unit labor costs will fall and the rate of inflation will gradually subside. Presumably the workability of moderate excess capacity rests on the idea that it can hold the basic wage settlement below the rate of growth of productivity. But with the strength of our labor movement rate of growth of productivity around 1 percent per year and inflation near 10 percent, it takes a lot more than moderate unemployment to bring this result about. The third solution is simply to have a recession or depression so large that the private cost-of-living escalator clauses are knocked out of the economy. Escalator clauses still exist, but firms quit honoring them just as Westinghouse quit honoring its uranium contracts. But this isn't a moderate policy. In a heavily indexed economy moderate policies either won't work or they will work so slowly that they are perceived as not working.

Moderate policies also depend upon the idea that the stochastic shocks which hit the system do not serve to offset the small positive effect of idle capacity. With rapidly rising oil prices, this is unlikely to be true. Even if the shocks are equal and opposite in value so that *in toto* we have an economy with no net good or bad luck, this does not mean that the impact of shocks is neutral. Because of the oligopolistic nature of our corporations and unions, our economy responds much more to upward price shocks than it does to downward price shocks. Upward price shocks get passed through quickly while downward price shocks get passed through slowly and tend to be absorbed in rising profit margins.

But the exogenous price shocks will also not have a zero-average value. When negative price shocks occur there will be pressures to stop these shocks from occurring through protection or price supports. If agricultural prices were to fall in international markets, we would simply institute more agricultural price supports. If the price of steel threatens to fall, we institute steel reference pricing. If the price of textiles threatens to decline, we increase protection for the textile industry. We will always be adopting more public policies that raise prices than reduce prices, just as we did in 1978. Where were the countervailing policies that should have lowered the cost-of-living index? They do not occur since they would have required someone's income to be lowered by a noticeable amount.

But there is another reason why moderation won't work. To create an economy with substantial amounts of idle capacity over a substantial period of time, it is necessary to be willing to tolerate a rising unemployment rate and slow growth.

Consider the problem of capital investment. With a substantial amount of idle capacity it does not make sense to invest in new facilities. But gradually, over time, slow growth plus depreciation eliminates whatever idle capacity exists. At this point the authorities either have to give up on their goal of idle capacity or further tighten monetary and fiscal policies to produce some more idle capacity. If they do the latter, they are once again depressing the rate of growth of the capital stock. If they accept the former, their anti-inflationary policy has withered away—at least as far as the anti-inflationary impact of idle capital is concerned.

On the labor side, a rising unemployment rate is necessary for two reasons. First, as the rate of growth of the capital stock is depressed so is the demand for labor. In the very long run, relative wages might fall to encourage the use of more labor-intensive techniques, but in the short run there are simply fewer jobs in the system. But there is also a dynamic within the labor market. Operating in a labor queue, employers always seek to get the best possible employees. In a period of high unemployment, employers try to replace less preferred workers with more preferred workers in the process of labor force turnover. The unemployment rates for the preferred workers fall and the unemployment rates for the less preferred workers rise even though the total level of unemployment has not changed. Over time this means that unemployment becomes more and more concentrated among the least preferred, and that the unemployment of the preferred groups gradually drops to what might be called full employment for that group.

But when the unemployment rate of the most preferred group reaches full employment, the policy makers once again have the same choice that they had in the capital area. They can further tighten policies to re-create unemployment among the preferred group or they can see their antiinflationary policy wither away. But the periodic tightening of policies and the resultant rising unemployment rate are a never-ending process. It continually has to be done.

This gradual rise in what is implicitly assumed to be full employment can be seen in the *Economic Report of the President*. In the Kennedy administration, 4 percent unemployment was set as an "interim" unemployment target because they did not want to defend even this level of unemployment. By the end of the Johnson administration, full employment was creeping up to 4.5 percent. By the end of the Ford administration, the economic report was defending 5 percent as full employment. In the 1979 economic report of President Carter, 6 percent is the implicit full-employment target. Given current policies, full employment will be even higher in the mid-1980s.

To some extent the problem is similar to our bombing policy of North Vietnam. We attempted to nibble them to death by gradually increasing the intensity of the bombing, but we only succeeded in making them more immune to bombing as we went along. In the end we dropped an enormous tonnage with little effect. The same tonnage dropped all at once at the beginning might have had a very different effect. The same with unemployment. To say that a dramatic sudden rise to some level, let's say 15 percent, would stop inflation is not to say that a gradual escalation to 15 percent over a number of years will stop inflation.

Miracle Cures That Don't Work

The other current solution is the idea that we can stop the problem if only we would balance the government budget. While there are many theoretical reasons why this won't work, it isn't necessary to go into these reasons to see the failure of the policy. Government budgets have come into balance in 1978 and 1979 at just the time inflation rates are accelerating. If balanced government budgets would stop inflation, it should have stopped.

At one time it was appropriate to look at the federal government to see what government was doing to the economy. Different levels of government were self-financing, and state and local governments always ran balanced budgets. With the growth of grants-in-aid, it is no longer possible to make a clear distinction between different levels of government. In the first three quarters of 1979, the federal government gave state and local governments $79 billion or 24 percent of what they spent.[20] During this time the federal government had a deficit of $9 billion while state and local governments had a surplus of $24 billion. Overall, government was running a surplus of $15 billion. The federal government could have cut its grants-in-aid by $24 billion, reduced the state and local surplus to zero, and given itself a surplus of $15 billion. Would that have made any difference to the economy? Presumably not since it is simply a matter of accounting where the government surplus appears. The only thing that affects the economy is the balance between what governments collect in taxes and what governments spend. While we could run a government surplus so large that we would produce the big bang recession and cure inflation, there is little reason to believe that a balanced government budget would cure the problem. That solution is in effect, but the problem is not being cured.

Our Fundamental Dilemma

Intractable problems are usually not intractable because there are no solutions, but because there are no solutions without severe side effects. Such is the case here. Each potential solution to the inflation problem lowers someone's income by a large amount; each increases someone's economic insecurity. It is only when we demand a solution with no costs that there are no solutions.

Chapter 4

Slow Economic Growth

INTEREST in accelerating economic growth has gone in and out of fashion. Along with the missile gap, it was one of the key campaign issues in 1960. The Russian growth rate exceeded that of the United States, and Nikita Krushchev was threatening to bury us economically and militarily. Faced with shortages of key materials and a sharp decline in America's productivity growth, accelerating economic growth has once again become an important issue.

The heart of the issue is productivity–output per man-hour. Our ability to consume ultimately depends upon our ability to produce. If we produce more per hour, each of us can have more purchasing power to buy the things we want. If productivity does not rise, our money incomes can rise, but it is not possible to have more real purchasing power. Often the issue is referred to as *supply-side economics*. How can we increase the supply of goods available for private consumption, corporate investment, and government expenditures? To find an answer we must find a way to accelerate the growth of productivity.

To stop inflation recent administrations have chosen to tighten monetary and fiscal policies to produce idle capacity. Whatever the merits of idle capacity in the fight against inflation, it exacts a stiff price in slower productivity growth. With idle capital, incentives to invest diminish. There is little need for new, more productive facilities. Knowing that they do not need to expand, firms often cut back on research and development for new production processes. With high unemployment workers fear that technical progress will cost them their jobs and that alternative work will be hard to find. Consequently they push for more restrictive work rules to stop technical progress. The end result is a stagnant economy with a productivity slowdown on top of a basic productivity growth rate that already puts us at the bottom of the industrial league–with about one-third the productivity growth of Japan.[1]

Unless this decline can be reversed, and unless productivity can be accelerated to the levels being achieved by West Germany and Japan, it is only a question of time until we slip into relative backwardness. Few major countries have been brought down by foreign enemies; many have disappeared because of their internal failures. How are we to eliminate our failures and make our economy more dynamic than it ever has been?

Here again the problem is not in finding policies that would significantly accelerate economic growth (there are many), but in adopting policies that would inevitably cause significant income reductions for someone. To increase investment someone's share of the national product must decline. Whose? Even more difficult is the process of disinvestment. We tend to think of economic growth in terms of investment and new products, but disinvestment is a necessary precondition. To have the labor and capital to move into new areas we must be able to withdraw labor and capital from old, low-productivity areas. But every disinvestment represents a threat to someone.

Disinvestment is what our economy does worst. Instead of adopting public policies to speed up the process of disinvestment, we act to slow it down with protection and subsidies for the inefficient. If our steel industry cannot compete, we protect it. If our television industry lags behind, we negotiate "orderly" marketing arrangements to keep out foreign-made sets. If textiles are a low-productivity industry that should be located abroad, we adopt stiff tariffs to preserve a local industry. Our shipbuilding industry is an industry completely dependent upon subsidies. All of these actions are designed to provide economic security for someone, yet each of them imprisons us in a low productivity area. If we cannot learn to disinvest, we cannot compete in the modern growth race.

The Process of Economic Growth

The process of economic growth can be compared to a complicated road-building operation. The first step is to scout the landscape and survey the terrain to see where you want to go and find the best possible routes for reaching desirable objectives. This is the role of scientific research. Generally scientific research proceeds far ahead of the rest of the road-building operation. We knew theoretically that an atomic bomb could be built four decades before we actually did it. At the moment, we know that fusion energy is theoretically possible (we can explode a hydrogen bomb), but several decades will have elapsed before we harness fusion reactions to generate electricity.

Well behind the frontiers of scientific research lies the domain of engineering research. The direction to go and the basic principles of how to get there are known, but a practical road must be designed. When engineering research has been completed, products and processes move from the domain of the theoretically possible into the domain of those processes that have been mastered and can actually be done. Using rockets for space travel was an idea whose origin is lost in the mists of history, but it passed the frontier of engineering knowledge when we were able to put a man on the moon and get him back.

While scientific explorations and engineering designs are both important, neither affects economic growth directly. The landscape may be known, the road can be built, but the road won't be built unless the economic benefits from having the road are greater than the costs of building it. Space travel is clearly feasible, but it costs so much that there is no economic demand for regular space travel to the moon. New knowledge only becomes relevant to our economy when costs have been reduced to the point that the information can produce goods and services, which we want, at a price we can afford to pay.

Further scientific and engineering research and development is necessary before a road will actually be built. Economic feasibility must be achieved. It is at this point that new knowledge starts to impact productivity. We build the road and start to use the new processes that produce better or cheaper products than we previously had. Our standard of living rises.

But an economy is not composed solely of new products and processes. It takes time and resources to shift to the new, so that any economy is a mixture of new, high-productivity activities and old, low-productivity activities. Some plants produce the newest products with the newest technologies while other plants produce old products with old techniques. The average level of productivity depends upon the relative weights in this mixture.

Our economy encompasses a wide range of productivities. Between broad industrial categories (see below) we had a productivity gap of almost five to one in 1977. Typically within each industry there is a range of productivities on the order of four to one. The result is a very wide distribution of productivities; but there comes a point when any product or process is so obsolete that it is no longer used. New products and processes drive old products and processes out of the economy. The old roads are torn up and abandoned.

This means that there are three factors that control the growth of productivity. First, how rapidly is the frontier of economic feasibility leading to higher-productivity activities. Second, how rapidly is the economy discarding low-productivity activities. And third, what is the distribution of activities between these extremes. Are most of our economic activities concentrated toward the high-productivity end of the spectrum or toward the low-productivity end of the spectrum? The frontiers of scientific and engineering knowledge are only relevant in that they are a distant road-building operation whose speed limits the long-run speed of movement toward higher-productive techniques and processes.

Already we are in a position to see some of the reasons why productivity has grown faster in countries such as Japan and West Germany. If a country is rebuilding from wartime devastation it will rebuild with new plants. Even if its best practice plants are no better than those in other countries, a larger proportion of the plants will actually be located near the best practice frontier. This will give them faster productivity growth, even if they have no advantage in terms of their best practice plants.

Often people talk of this phenomenon as if it were better to lose a war and have your country blown up than to win a war and escape destruction. This is simply silly. To recover, the West Germans and the Japanese must devote a large fraction of their GNP to investment. The result is a much lower standard of living during the recovery period. Productivity grows rapidly but only at the sacrifice of real standards of living. If it were an advantage to have your country blown up, the winners could do likewise. They could junk their old plants (bomb them if you like) and build new plants. They don't because to do so is to reduce their standard of living. Consumption would have to fall both because production is down and because investment would have to rise.

Rebuilding countries do, however, have an advantage. Often countries find it difficult to get out of low-productivity industries and products even when economic analysis would call for it. Individuals lose their jobs and firms go bankrupt. Workers and firms lobby for government protection, subsidies, and regulations. If they are successful, the economy is locked into low-productivity operations much longer than economic circumstances would warrant. In the devastations of a postwar period, there is nothing to protect or subsidize, and no one could afford to do so even if there were. The economic losses that have been suffered can be blamed on someone else's army. The net result is that obsolete industries are not rebuilt. Disinvestment in low-productivity industries occurs at a much faster rate than it usually does.

Disinvestment

While there are many voices calling for more investment, the process of disinvestment is even more important. Eliminating a low-productivity plant raises productivity just as much as opening a high-productivity plant. But doing so takes fewer resources. Large investments are not necessary. To close a low-productivity plant also makes it possible to move the workers and capital that have been tied up in this activity into new, high-productivity activities. With more men and investment funds, new activities can grow more rapidly. Paradoxically the essence of investment is disinvestment.

While we may have problems with research and development and with investment, our main failure lies in the area of disinvestment. We simply are not very good at accomplishing it. This is one of the places where the mixed economy has not worked. Capitalism is, after all, a doctrine of failure. The inefficient (the majority) are to be driven out of business by the efficient (the minority), and in the process productivity rises. Yet we are extremely reluctant to practice this part of our economic religion. This reluctance has a real moral basis at the level of the individual (a failing individual is a starving individual), but it has no moral basis when it comes to firms. Yet if anything, we have more programs to protect institutions (all of course justified in the name of protecting individuals) than we do individuals.

Low-productivity firms are often located in industries where the demand is stagnant or falling. This is partly due to the fact that new plants do not need to be built to meet new demands, but it is also due to a human problem. Dying industries simply cannot be managed as efficiently as growing industries. Growing industries attract bright aggressive managers who want to advance rapidly with their companies. In dying industries promotions are few and far between. Smart young managers know that they should be avoided. Who wants a job where the basic problem is to decide who to fire each day and where new, exciting investments are not happening? In a dying industry everyone is out to protect what they have rather than to build something better. They know that any gains in efficiency will simply result in more layoffs.

The result is a set of attitudes and actions on the part of both managers and workers that make it virtually impossible to have rapid productivity growth in an industry where output is not growing or falling. The phenomenon can be seen all across America from railroads to schools. Efficiency falls as output drops. In the Boston area, where school enrollments are now rapidly falling, I know of no school system anywhere that has managed to reduce personnel nearly as fast as enrollments have fallen.

The basic problem in disinvestment is the desire each of us has to avoid the economic pains that are endemic whenever disinvestment occurs. Someone is worse off because of those disinvestments, and they have every incentive to appeal for government aid to stop or slow down the process of disinvestment. Regulations are adopted to stop railroads from abandoning noneconomic lines. Subsidies are used to keep an inefficient shipbuilding industry in business. Instead of shrinking with declining enrollments, schools discover special education and the need for more teachers. While it is easy to say that such things should not occur, each of us would be demanding the same protection if we were in the affected industries or communities.

Process Innovations

Often the productivity problem is portrayed as if it were a simple problem of too little investment. If we just cut consumption and invested a larger fraction of our GNP, our productivity would be higher. One of the problems with this analysis is that more investment would now be occurring if it were profitable to do so. In most of the post–World War II period our economy has had the problem of wanting to save more than it wanted to invest. The result has been a series of recessions where demand (consumption plus investment) was below what the economy could produce. If anyone had wanted to invest more, there was no shortage of savings or production facilities, yet the investment did not occur. Taxes are often blamed, but our business taxes are no higher than those abroad. For some reason we just do not seem to have as many profitable investments.

Part of the explanation for this lack of investment can be seen in the context of what economists call *learning curves.* The learning curve phenomenon first came into focus in the production of Liberty ships and airplanes during World War II. After the plants were built and in operation, the number of man-hours of work necessary to build a ship or airplane fell rapidly as more and more ships or airplanes were built. The capital equipment did not change appreciably, but productivity rose dramatically.

This same phenomenon has been widely observed in civilian production. Following the introduction of a new product or the start-up of a new plant, labor costs typically drop sharply for a few years and then more slowly, even though the labor force is working with the same capital equipment. Investments ultimately prove to be profitable or unprofitable depending upon the steepness of the learning curve and the pace of productivity advancement after the plant has been built. Based on engineering data, it is not easy to predict production costs since production costs are not constant over time. Multinational firms find that they can build the same plant in different countries or different regions and yet have very different productivity results.

The learning curve is related to the process of informal, on-the-job acquisition of skills and team productivity.[2] In the process of production workers learn and improve their individual job skills and learn to work together as a team. New workers are inferior to those that have been working on the job for some period of time, even though their formal education and skills may be identical. As a product is being built, new and better ways of building it are found with experience. Each innovation in the production process may be small, but the cumulative effect of many small improvements is often large.

The net result is a sharp rise in productivity as a plant goes down its learning curve. Labor costs of production at the bottom of the learning curve are often a mere fraction of those at the top. But the process is not automatic. It depends upon high quality management and a cooperative work force. If the work force is unhappy, it can stifle the learning process. If managers are incompetent, opportunities for new labor-saving procedures are missed. An early adoption of rigid work rules can freeze the plant into its initial productivity level and prevent it from proceeding down the learning curve. (This is an important factor in Britain where rigid work rules are usually negotiated before a plant goes into production.) The problem is to descend as far as possible and as quickly as possible down the learning curve. The firm that does so will have the lowest costs of production and the most profits.

This creates an interdependence between capital and labor that is not recognized in the simple cry to raise investment. If the Japanese are able to generate a steeper learning curve than Americans, the same steel mill may be a good investment in Japan and a poor investment in the United States. To raise investment it is necessary to improve the characteristics of the labor market. New skills and higher earning depend upon new investments, but new investments also depend upon a cooperative work force. Simply raising the income of capitalists, with tax cuts that must be paid for with tax increases for workers, is unlikely to achieve either more investment or a higher growth of productivity. In generating more profitable investment opportunities, skill acquisition and a cooperative work force are as important as more funds to buy new equipment. Starting a class war is hardly the way to proceed. Imagine what those who believe that all work effort is dependent upon large income differences would predict about an economy where large firms give lifetime jobs, where relative wages are almost completely dependent upon seniority rather than personal skills and merit, and where income differentials are 50 percent smaller than in the United States. Yet the Japanese have the world's highest rate of productivity growth. *Why?*

The answer is found in the incentives this system provides for going down the learning curve. With lifetime employment and seniority wages, technical progress is not threatening. Whatever is invented, it is not going to threaten either employment or wages. With the typical worker getting about 50 percent of his or her wages in twice yearly bonuses that depend upon profits, a steep learning curve is of direct concern to each worker. Every worker has an incentive to maximize productivity by welcoming technical change, learning new skills, and contributing to industrial teamwork in a way that makes U.S. employers envious. Often this phenomenon is dismissed as a cultural one impossible to duplicate in the United States, but it probably has more to do with the economic incentive system than it has to do with culture. Faced with the same incentives, U.S. workers would respond in the same way. In any case, we need to find some system that achieves the same results.

Recent Declines in Productivity

While we need to do much more than simply reverse our recent slowdown in productivity, the slowdown is interesting since it sheds some light on what might be done to accelerate productivity. But more importantly, it vividly illustrates the complexity of the problem and the irrelevance of simple one-factor solutions such as more investment.

There is no doubt that the rate of introduction of new products and new processes has fallen. Productivity in the private business economy was growing at 3.2 percent per year from 1948 to 1965, at 2.3 percent per year from 1965 to 1975, and at 1.1 percent from 1972 to 1978.[3]

A wide variety of possible causes has been suggested for the lack of performance. Research and development expenditures are lower now than they were in the 1960s. It is often said that investment has fallen. We invest a smaller fraction of our GNP in plant and equipment than most of our industrialized neighbors. Government health, safety, and environmental regulations may have made growth more difficult. The age-sex mix of the labor force has been shifting toward inexperienced (low-productivity?) workers—women and the young. Stop-go economic policies and inflation have made investors reluctant to invest. Uncertainty has risen. Workers are alienated and less cooperative in producing productivity gains. With high unemployment and more fears about job security, work rules have become more restrictive. The list of possibilities is almost endless.

Two of the commonly suggested causes simply do not fit the facts. Research and development expenditures are down from 3 percent of the GNP at the beginning of the 1970s to slightly more than 2 percent at the end of the 1970s, but productivity started to fall in 1965 well before the downturn in R&D expenditures.[4] In addition, as we have seen there is a long-time lag between R&D and productivity. A lack of R&D in the 1970s may cause productivity problems in the 1980s, but it does not explain productivity problems in the 1970s much less than in the 1960s. Our industrial neighbors have also consistently invested less in R&D than we.

Plant and equipment investment cannot explain the decline because it is up, not down. When our productivity was growing most rapidly (1948–65), plant and equipment investment averaged 9.5 percent of the GNP. Productivity growth fell after 1965, but investment rose to 10.2 percent of the GNP from 1966 to 1972. Productivity growth took another fall after 1972, but investment stayed up at 10.1 percent of the GNP from 1973 through 1978, despite the sharpest post–World War II recession.[5] Perhaps we should invest more, but declining investment is not the source of our problems.

If you analyze the pattern of productivity growth, it is clear that large productivity gains are associated with any surge to full employment. Conversely productivity growth falls as the economy moves away from the full utilization of men and machines. This occurs because we have a large proportion of overhead labor and plants designed to operate most efficiently at capacity. Managers, research departments, salesmen, maintenance workers, and the like either cannot be or are not cut back proportionally when output falls. The result is a drop in productivity since more man-hours are now necessary to produce a unit of output. Conversely when output rises toward capacity, we do not have to expand the overhead labor force. Output goes up, but overhead man-hours do not go up, and the result is a rapid gain in productivity.

About 30 percent of our productivity slowdown can be attributed to idle capacity. In our efforts to fight inflation, we have deliberately chosen to hold the demand for goods and services below what the economy could produce. Whatever benefits this may create in terms of less inflation, one of the costs is a slower rate of productivity growth. This part of the productivity problem will only be cured when we solve the inflation problem or decide to fight inflation with some other technique.

About 40 percent of the decline can be traced to a shift in the mix of goods and services being demanded and produced. If there are substantial differences in productivity between industries, as there are, the mix of output demanded by consumers, business, and government can have a substantial effect on the rate of growth of productivity. If demands are shifting toward high-productivity industries, the economy's productivity will grow rapidly. If demands are shifting toward low-productivity industries, the economy's productivity will grow slowly.

In the United States there are large differences in productivity between industries. In 1977 a man-hour of work produced $4.92 (1972 $) worth of output in services and $23.59 worth of output in finance. This is a range of almost five to one.[6] Despite what is often believed, manufacturing productivity ($8.44 per hour in nondurables and $8.42 in durables) is not much above that of the economy as a whole ($8.09). High productivity industries are finance, wholesale trade, utilities, communications, and mining. Low productivity industries include services, retail trade, construction, and agriculture.

With such wide differences in productivity, the mix of goods and services demanded can have a large effect on productivity. For a substantial period of time after World War II, the mix effect was enhancing productivity. We were leaving low-productivity industries, mainly agriculture, and entering high-productivity areas. But this process ended around 1972. The mix of goods and services demanded started to decelerate the rate of growth productivity rather than accelerate it. The sharp movements out of agriculture ended, and services (another low-productivity industry) started to grow much more rapidly.

From 1948 to 1972 agriculture, an industry whose productivity was 60 percent below the national average in 1948, reduced its demand for labor by 500 million man-hours per year. Every worker leaving agriculture and entering the urban economy meant a sharp rise in productivity, and there were millions of such workers. But by 1972 this process had essentially ended. Productivity was still rising rapidly in agriculture, but agriculture had become so small that it no longer was releasing millions of workers. After 1972 annual reductions were down to 50 million man-hours per year. Large amounts of labor were no longer being released from a low-productivity industry.

Quite the reverse was now occurring. Another low-productivity industry, services, started to grow much more rapidly. While less than 30 percent of the additional man-hours added to the economy from 1965 to 1972 had been in services, 47 percent of all man-hours added to the private economy after 1972 were in services. Since service productivity is 40 percent below the national average, every worker moving into services represented a sharp cut in average productivity. What had been a sharp shift toward higher productivity became a sharp shift toward lower productivity.

Almost half of those extra services workers went into health care. If we want health care that is what we want, but one of the inevitable consequences is a lower growth of productivity. The essence of the problem can be seen in the three hundred thousand security guards added to our economy since 1972. Since security guards protect old goods and do not produce new goods they add nothing to output, but they increase man-hours of work. The same number of passengers are flying from Boston to Los Angeles, but now it takes more hours of work to get them there since their luggage must be checked. The net result is a decline in productivity even though our sense of well-being may be up.

The remaining 30 percent of the decline can be traced to particular problems in three industries—mining, construction, and utilities. Mining and construction have even experienced negative productivity growth. Output per man-hour is less now than it was a decade ago. Utility productivity growth is down sharply.

The decline in productivity growth in electrical, gas, and sanitary utilities is the easiest to explain. This is a clear case where productivity growth is highly dependent upon the growth of output. Additional output is produced in new efficient plants, and a very large fraction of the labor force is overhead labor needed to maintain the distribution systems. When more energy is consumed, output goes up very rapidly relative to employment. Conversely when output stabilizes or goes down, productivity stabilizes or goes down. With the much higher prices of energy, output growth has slowed and some years even fallen, with a sharp fall in productivity growth from over 6 percent per year to 1 percent per year. The obvious cure is a return to rapidly growing consumption, but this is not likely given what is expected to happen to energy prices.

Mining productivity has fallen 23 percent since 1971. This is the one place where it is possible to lay part of the blame at the door of new health, safety, and environmental regulations imposed by government. But much of the problem is due to geology. Less oil is being produced from many more wells, and this shows up as a decline in productivity.

This is not to say that the regulations are either unwise or unwarranted. Greater health, safety, and environmental protection simply impose large costs in mining. If we want safe mines and a clean environment, we are going to have a slower growth of productivity in mining, at least for awhile, than if we do not want those things.

Construction is another industry with negative productivity—down 19 percent since 1968. Here the problem is partly a measurement problem and partly a real problem. How do you measure output in an industry that does not produce a standardized product? The standard technique is to use inputs (the volume of construction materials) to measure output, and this may underestimate real output if progress is being made in using materials more efficiently. We may also be demanding more variety in our construction—fewer large housing projects, fewer massive road projects—than in the past. And as a result, the construction industry does not get to take advantage of its learning curve or economies of scale.

As the decline in productivity is examined more closely, simplicity disappears. Even when the causes are clear, the solution is not. We have elected to fight inflation with idle capacity, and this explains 30 percent of our productivity decline. If agriculture were still disgorging massive amounts of labor, our productivity would be higher, but that phase of our industrial life is now over. Productivity would be higher if we did not want so many services, but the demand for services is only a problem if we were in some sense buying more services than we really want. One can argue, for example, that medical insurance leads us to buy more health care than we would buy if insurance were not available and every bill had to be paid in cash. But who wants to go without health insurance? Our real standard of living would grow more rapidly if we happened to want goods made in high-productivity industries, but we don't. To buy a high-productivity good that you do not want is not to raise your real standard of living, although it would accelerate the growth of productivity.

Productivity growth would be higher if energy prices were falling and consumption was rising, if it were easy to make mines safe and environmentally sound, and if we all wanted to live in identical homes, work in identical factories, and shop in identical stores. But none of these things is possible. Simply raising investment might lower productivity growth since it would allocate more resources to an industry with below-average productivity—construction.

Just as the causes of our productivity slowdown are complex and varied, so will be the cures. If you remember that productivity has been growing at about 3 percent per year for as long as we have been measuring productivity growth (well back into the nineteenth century), and that our neighbors have achieved growth rates double or triple this in the last few decades, it is very unlikely that there will be a simple cure. Current productivity growth rates are deeply embedded in the structure of our economy, and major changes will be necessary before we see major improvements.

Productivity and International Competition

While advance in productivity in any sector contributes to our overall standard of living, our international competitiveness is primarily dependent upon what happens to productivity in two industries—agriculture and manufacturing. In this we differ from most other industrial powers in that over 20 percent of our exports are agricultural commodities. While we lag in manufacturing, in agriculture we continue to lead the productivity race with more than a 6 percent annual gain.

In agriculture the problem is not productivity but opening foreign markets to our producers. Agriculture is the industry that everyone, including ourselves, protects the most. For all practical purposes, the United States is a residual supplier to the rest of the world. Each country buys only what it cannot produce itself. Operating behind high price supports, Common Market farmers produce whatever they can. If crops are bad, the Common Market is a massive agricultural importer. If crops are normal, the Common Market is a large importer. If crops are very good, the Common Market subsidizes exports. Other countries do the same. This leaves us subject to large demand shocks and sudden price changes, but it also deprives us of one of our major export markets. As a consequence, we become more dependent upon our relatively weak sector—manufacturing.

To survive in today's international competition we must push for freer trade in agricultural products. It is our area of greatest comparative advantage. But it is also an area that illustrates our basic problem. While we have a large comparative advantage in the production of most agricultural commodities, we are not in a position to push for free trade since we protect weak agricultural areas (sugar, cheese, and processed meats) as much, or more, than the rest of the world protects their farmers. We need free trade in agricultural commodities if our economy is to compete, but we cannot demand it because we do not practice it. Overall, farmers would make large income gains, but particular farmers in some regions of the United States would lose. Here again, we cannot play an economic game with a substantial zero-sum element.

To keep pace economically we are going to have to give up our own protection in some areas and demand access for our products in other areas. We no longer can afford to accept a world where our agricultural commodities are excluded. Basically this is going to mean getting tough with our allies. West Germany and the rest of the Common Market have got to stop preaching free trade in manufacturers while practicing protection in agriculture. If Japan wants to export cars it has to be willing to import food.

Very limited progress was made in the recent Tokyo round of trade negotiations, but it was so limited as to not even constitute the first step in a very long march. Future trade negotiations must make progress on agricultural commodities. If necessary, we should being limiting others's manufacturing access to our markets if they do not give us agricultural access to their markets.

In manufacturing there is no evidence of a slowdown in productivity once a correction is made for idle capacity, but this is the area where our productivity growth rates are the poorest relative to the rest of the world. Often these problems are blamed on American multinational corporations. As is true in most cases, the worries are greatest when the problems have past their peak. In the past, United States multinationals undoubtedly moved production abroad faster than would have happened if they could not have owned those foreign facilities. But this activity is clearly on the decline.

Multinationals need low wages, stable governments, and educated labor forces to establish facilities that can compete with those in the United States. In Europe low wages are already gone. And they are rapidly disappearing in those parts of Asia (Korea, Taiwan, Hong Kong, Singapore) that have stable governments and an educated labor force. There are, of course, many countries with low wages that will be attractive for low-productivity industries, but these are precisely the activities where we should be disinvesting. If anything, foreign multinationals should contribute to manufacturing productivity in the future. Since wages are lower here, they are now starting to enter the United States. In the past we had little to gain, but now a company like Michelin brings us knowledge about producing radial tires that we do not seem to possess. When foreign multinationals enter the United States they speed up the transmission of industrial knowledge from high-productivity areas abroad to low-productivity areas in the United States.

Accelerating Productivity

Outside of agriculture, our basic problem is accelerating the growth of productivity. Our research and development expenditures may be too small (the right proportion of GNP to devote to R&D is one of those imponderables), but the real problem is a substantial bias toward developing new products rather than new processes for producing old products. This bias exists for two reasons.

First, new products are always more glamorous than new processes for producing old products. Scientists and engineers would rather have government R&D money go into new products. Second and more importantly, we have great difficulty in publicly funding process R&D in an economy where production is almost always in the private sector. When government funds are used to finance the development of new products in universities, no one can predict the chief economic beneficiary if success is achieved. Government officials cannot be accused of deliberately raising the income of some particular firm. In process R&D, however, the potential gainer is clearly identifiable—the firms that now make the product in question. In the case of defense or space we are willing to provide public R&D funds for process improvements since government is the ultimate buyer of the products that will be more efficiently produced. But when it comes to civilian production, we are reluctant to provide public funds for process R&D, since the question arises as to why the taxpayer should have to contribute to make some stockholder richer. To engage in process R&D, tax money must be taken from one private individual and given to another private individual. But this is what we cannot do.

Yet, as we have seen, the heart of the productivity problem lies in quickly advancing down the learning curve. Process R&D expenditures are needed to generate a steep learning curve, but the learning curve lies in the private sector. One could argue that the private sector should finance its own learning curves, but there are good reasons why R&D is financed by government both here and abroad.

R&D expenditures are financed by government for the simple reason that no private firm can hope to appropriate all of the benefits that might occur. A new product may be developed, but it may not be of use to the firm financing the work. The firm does not have the expertise or complementary products necessary to take advantage of the breakthrough. If the product is developed with public funds, the research lies in the public domain, and the firms that can use the product can gain access to the knowledge necessary to exploit it. Governments pay for R&D since what may be a good investment for the whole society may be so risky for any one firm that it will not undertake the expenditure.

The same inability to appropriate all of the benefits exists with process R&D. Suppose a new product has been developed—the solar cells that directly transform sunlight into electricity for our existing space satellites. Given current volumes and production techniques, they are too expensive for earthly electricity generation. You are a manufacturer thinking about civilian production. You know that large process R&D expenditures and large initial investments would be necessary to go into production. If the learning curve is very steep your investment will be profitable, but a shallow learning curve could exist which would make your investment unprofitable.

If you were certain that you and only you could reap the benefits of success, you might take the risk, but you know that this is very unlikely. If you succeed and the learning curve is very steep, you may get your product to market first, but other firms now know that success is possible. They can start production knowing that a steep learning curve exists. Eventually they will find the path you found and gain some of the benefits you were counting on.

In essence the problem is similar of that of a book on chess end games. If you are told that the game can be won in four moves, it is almost always possible to find the four moves, but in a real game not knowing that victory is within your grasp you do not look hard enough and never find the four moves. The first person down the learning curve paves the way for the followers. He demonstrates that success is possible. But not being able to get all of the benefits, no one may be willing to be first. And even if someone is willing to be first, we have an inefficient process where different firms must essentially reinvent the wheel—the desired process.

To speed up productivity we must find an acceptable technique for involving government in process R&D. There are probably three essential ingredients. First, we all have to accept the fact that any government program is going to help someone. The fact that the winners can be identified ahead of time does not make a program wrong. As long as we have a private profit system of enterprise, any public efforts to raise productivity will make more profits for someone. Second, there is nothing wrong with profits and making someone rich if we have a fair system of taxation. Tax reform, and having what is perceived as a fair tax system, is an important ingredient in stimulating productivity since it allows us politically to shift R&D to areas where it can have a large productivity payoff. Third, process innovations paid for with government funds should be available to everyone in an industry. Other firms should be able to study how it was done at the first firm so that they can get the same productivity gains without having to make duplicating R&D expenditures.

This is essentially what we now do in agricultural R&D where new processes are tried out on experimental farms and then offered to all farmers. In this case the experimental farms are owned by government, but this is not possible in most industrial operations where production units may be very large. But however we do it, we must restructure the economy so that we can engage in more process R&D and gain more of the potential benefits of steep learning curves. New products are important, but at any point in time most of the economy is composed of old products. Making these products more efficiently is the heart of the productivity problem.

While foreign analogies should be treated with caution, it is instructive to think about Japan's success with process innovations. They have not been a leader in new products, but they have been a leader in better processes for producing old products. This springs from the absence of a sharp dividing line between public and private, and a willingness to engage in process R&D. But to do this the Japanese must take revenue away from some Japanese and give it to other Japanese. We are reluctant to do this when it comes to private corporations because we do not trust them to give the benefits back to us, and because we cannot justify a transfer of resources from one American to another.[7]

Accelerating disinvestment is the second ingredient in speeding productivity growth. Ending subsidies, protection, and favorable regulations will help, but we are not going to be able to do so until we find a way to provide economic security for individuals without providing economic security for failing institutions (see chapter 8). At the same time we also need to go beyond a free market policy that promotes disinvestment and encourages reinvestment in high-productivity areas.

We do not need central economic planning in the sense of an agency that tries to make all economic decisions, but we do need the national equivalent of a corporate investment committee to redirect investment flows from our "sunset" industries to our "sunrise" industries. Such committees play an important role in the investment decisions of large corporations, and they could play an equally important role in national investment and disinvestment decisions.

With our current system of internal finance, growth in high-productivity areas is often limited by the funds that can be internally generated. This often lowers their growth and our national growth below what it should be. Similarly with internal financing, "sunset" industries often have access to plentiful funds for new investment. They can reinvest their internal savings, but their steady cash flows also let them borrow in the capital markets. Often these investments should not occur. A national investment committee could help make sure they did not occur.

For most of our industrial competitors the central bank plays an important role in allocating investment funds. In addition to worrying about the money supply and the rate of interest, it attempts to direct funds toward areas of major national interest. The system is probably most heavily developed in Japan but exists to some extent in Italy, France, and West Germany. In the past our Reconstruction Finance Corporation played a similar role. It could and did provide major funding for large projects in new areas.

A national investment bank could be regarded as a competitor with private banks or it could work through private banks as it does in Japan. It certainly represents more government in the mixed economy, but the time has come to recognize that if we are going to compete with some of our more successful industrial neighbors, we are going to have to change the way we have been doing things in the past. Simply retreating into the past and calling for the end of government involvement won't solve the problems. We have to do much better than we have ever done pre- or post-New Deal if we are to compete in the productivity race of the 1980s.

While there is much to be gained by taking our foot off the current economic brakes on economic change, we must also learn to put our foot on the economic accelerator. If others have learned how to more quickly reorientate their economy to new growth areas, so must we. We do not have to reinvent the wheel, we merely have to adopt and adapt what others have learned to our culture and institutions.

While a decline in investment did not cause our current productivity problems, an increase in investment is probably one of the ingredients in a cure. Those who are doing better than we invest a substantially larger fraction of their GNP. But to invest more, we have to do two things. We have to create incentives to increase investment, and we have to accumulate the necessary funds for this investment.

The simplest part of the problem is increasing the incentives to invest. The easiest solutions would be to abolish the corporate income tax and integrate corporate and personal taxation. With full integration, there would be no corporate income tax but each individual shareholder would be liable to pay personal taxes on all income (retained or paid out) earned on his or her behalf. At the end of the year, shareholders would get the equivalent of a W-2 form telling them how much income to add to their other sources of income and how much income tax had been withheld on their behalf.

Since corporate after-tax rates of return would approximately double, corporate managers would have a strong incentive to increase investments. At the same time, we would increase both the equity and progressivity of the personal income tax. Each shareholder, rich or poor, would now pay taxes at a rate commensurate with his own income position rather than at some common rate. Taxes would go down for some; up for others.

The corporate income tax should be abolished regardless of whether you are a conservative or a liberal. Based on our principles of taxation, the corporate income tax is both unfair and inefficient. In a country with a progressive personal income tax, every taxpayer with the same income should pay the same tax (horizontal equity), and the effective tax rate should rise in accordance with whatever degree of progressivity has been established by the political process (vertical equity). The corporate income tax violates both of these canons of equity. Consider the earnings that are retained in the corporation on behalf of the individual stockholder. Low-income shareholders with personal-tax rates below the corporate rate of 46 percent are being taxed too much on their share of corporate income. To the low-income shareholder the corporate income tax is unjustly high. Conversely, high-income shareholders with personal-tax rates above 46 percent are being taxed too little on their share of corporate income. To the high-income shareholder the corporate income tax is a tax shelter or tax loophole. As a consequence, vertical equity is being violated. Horizontal equity is also being violated, since two individuals with exactly the same income will pay different taxes, depending upon the extent to which their income comes from corporate sources.

It is important to notice, however, that to eliminate the horizontal and vertical inequity of the corporate income tax, the tax must be eliminated on both dividends and retained earnings. Simply eliminating the corporate income tax on dividends increases the tax shelter aspects of the tax without achieving equity.

While corporations are legal entities that write checks to government, they do not pay taxes. They simply collect money from someone—their shareholders, their customers, or their employees—and transfer it to government. There is no such thing as taxing corporations as opposed to individuals. This immediately raises the issue of who ultimately pays the corporate income tax. The incidence of the corporate income tax is an area of economics with a large literature and little or no agreement. Depending upon the exact assumptions used, the definition of incidence, and the time periods under consideration, it could be a tax on shareholders, a sales tax on consumers, or a tax on employees. (Personally, I believe that it is a tax on shareholders in the short run and a sales tax in the long run, but my advocacy of its elimination does not hang on that belief.) While there may be a certain perverse political virtue in collecting a tax where no one is sure whether he pays it, simple economic efficiency and equity would seem to call for the elimination of taxes where incidence is uncertain. Only if we do so can we establish a tax system that is fair and has the economic consequences we intend.

Since interest payments are deductible business expenses while dividends are not, the corporate income tax also biases the structure of capital toward debt capital and away from equity capital. Debt capital becomes cheaper than equity capital, not because that is true in the market, but because the tax laws make it so. From the point of view of the efficient allocation of capital and an efficient capital structure, there is no reason why government should be intervening to bias business choices in the direction of debt capital and away from equity capital. From the point of view of having a healthy, vital capitalistic economy, government should, if anything, be doing the opposite. Eliminating the corporate income tax would eliminate this bias in capital structure and hence improve the efficiency of the capital market.

But the capital market would also be improved in another way. Since the maximum personal tax rate (70 percent) on property income is substantially above the corporate income tax rate (46 percent), and most corporate shares are held by high-income individuals, there is a strong incentive for firms to retain earnings, reinvest them, and provide benefits to their high-income shareholders in the form of a larger capital stock and higher stock prices. As long as the stock is held, no personal income tax will be paid, and when it is sold, only the lower capital gains tax need be paid. While there is nothing wrong with retained earnings, it once again should be up to the market rather than the tax laws to determine how much income should be retained rather than paid out to the shareholders. Eliminating the corporate income tax would remove the tax incentive to retain earnings. As a result, both the supply and demand for funds in the capital market would increase, once again leading to greater efficiency.

To the extent that the corporate income tax is in fact a sales tax collected from the buyers of corporate products, a number of benefits would accrue from its elimination. The prices of corporate products would gradually fall with favorable effects on the rate of inflation. Because of lower prices, our competitiveness in international markets would also increase, and this would be especially true vis-à-vis countries that can rebate their value-added taxes on exports. The net result would be more goods sold and more Americans employed.

If all of these advantages exist, why do we have the corporate income tax, and why is it still defended? Many people, including the man on the street, think that it is a way to tax the rich. As I have shown, this is simply a mistaken perception. To the extent that the corporate income tax is a sales tax or a tax on employees, it is not a tax on the rich. Even if it is ultimately paid by the shareholder, it is not a very good tax on the rich. To tax the moderately rich we must tax the poor at very high rates and provide a tax shelter to the very rich. If we want to tax the rich, the personal income tax is the right way to do it.

Some liberals oppose the elimination of the corporate income tax on the grounds that low-income stockholders would not have enough cash income to pay the taxes owed on the earnings retained on their behalf. This is not a problem since corporations could be required to withhold taxes for shareholders just as they now do for employees. Every year shareholders would receive the equivalent of a W-2 form that would list their corporate earnings and the taxes that had been withheld on their behalf. Those overwithheld would receive a refund, and those underwithheld would have to pay the additional taxes owed just as they now do on their wage and salary earnings.

Some business managers support the corporate income tax on the grounds that it encourages retained earnings, and it gives them more funds not subject to the competitive bidding of other managers in the capital market. To some extent this perception is undoubtedly true but I suspect that shareholders would still be willing to tolerate some substantial amount of retained earnings in a system where the tax system was neutral with respect to whether earnings were or were not paid out.

Those who manage government often oppose the taxation of corporate income as personal income on the pragmatic grounds that less revenue would be collected and thus some other tax would have to be raised. Depending on exactly which estimate of the distribution of stock ownership by income class is used, Treasury losses range from $4 to $10 billion or from 2 to 5 percent of the revenue now collected from personal and corporate income taxes. To put this amount in perspective, simple elimination of the corporate income tax on dividends would cost the Treasury $13 billion. The revenue shortfall arises not so much because individual shareholders would pay less than they now do (some would pay more, some would pay less, and the balance depends upon the distribution of stock ownership by income class), but because a substantial amount of stock is owned by institutions (charities, pension funds, and so forth) that do not pay personal income taxes.

In the long run much of this shortfall would be recouped, and that which is not recouped would yield substantial benefits. To the extent that pension funds have higher incomes, they are either going to reduce contributions (leading to higher taxable incomes) or increase the pensions paid (leading to higher taxable incomes). If nonprofit charities have higher incomes, the public will, to some extent, give less to charities (leading to higher taxable incomes). To the extent that the higher earnings of charities are not offset by lower annual giving, they will be doing more good works. And this is, after all, why we made them tax exempt in the first place. If we really want to tax them, we can easily pass a law doing so in any case. At the moment we are simply being inconsistent and taxing their corporate, but not other, sources of income.

When you review the arguments, there isn't any case for the retention of the corporate income tax. It is both unfair and inefficient. It ought to be eliminated. And all corporate incomes–retained or paid out as dividends–ought to be taxed at personal income tax rates appropriate to the shareholders who own them. In doing so we will increase the fairness of the tax system, improve the allocation of investment funds, and create a powerful incentive for more investment.

Increasing the incentives to invest is relatively simple, but raising the necessary funds for investment is difficult–not economically but politically. We are confronted with the question that I posed to the Harvard alumni reunion. If we were to raise investment from 10 percent of the GNP to the 15 percent level of West Germany or to the 20 percent level of Japan, who would be willing to give up 5 or 10 percent of the GNP? Conservatives say that we should generate the extra savings by lowering taxes on savers and raising taxes on consumers. Basically this means shifting the tax burden from rich to poor since savings propensities are naturally much higher for the rich than for the poor.

There is no doubt that the extra savings could be raised in such a manner if the shift in the distribution of income were sharp enough. Suppose that households with incomes below $16,000 per year (the bottom 60 percent of the population in 1977) saved nothing, and that households with incomes above $38,000 (the top 5 percent of the population) saved 50 percent of their extra income.[8] To raise savings by 5 percent of GNP you would have to transfer $188 billion from the bottom 60 percent of the population to the top 5 percent of the population. This would lower the real standard of living of the bottom 60 percent of the population by 25 percent.

The income of the top 5 percent would rise by 46 percent. (In fact the transfers would probably have to be larger than this since the bottom 60 percent do some saving and the top 5 percent may not have a 50 percent marginal savings rate.) To accomplish the necessary objective—more savings—a majority of the population would have to endure a sharp reduction in their current consumption. Not surprisingly they are reluctant to do so. Yet more savings are necessary if more investments are to be made.

The direct way to solve the problem in an equitable manner is simply to run a surplus in the government budget of the appropriate magnitude. Taxes are raised by the necessary amount, and each of our incomes is reduced in accordance with a tax system. If we have an equitable tax system we have an equitable spreading of the burdens. But this directly poses the question of what is an equitable tax system and an equitable distribution of after-tax income.

More investment, speedier disinvestment, more process R&D—they all pose the fundamental zero-sum distributional question. Someone's income will have to go down and these losses are going to be substantial. For those that lose, the existence of even larger social gains are irrelevant. They are only interested in preventing their losses.

Chapter 5

Environmental Problems

PART of the reason why we don't seem able to compete in the growth race is that we are not sure that we want to compete—"small is beautiful." In one corner of the political arena we have those who want to restructure the economy to stimulate growth; in the other corner we have those that want to restructure the economy to limit growth. From the latter perspective zero economic growth (ZEG), if only it could be achieved, would result in the "good" society. Natural resources would be exhausted less rapidly, pollution would be less intense, and everyone would be happier in a society where we weren't all struggling to have more.

Technically the quarrel could easily be resolved since the two groups are not really in opposition. ZEG advocates want high productivity growth with low output growth. With higher productivity, fewer inputs (especially nonrenewable inputs) are needed per unit of output. New processes make it possible to cut the amount of pollution associated with any level of output. Similarly those who want more growth are really interested in productivity. If our productivity is high, we can compete regardless of how many goods and services we choose to consume. If Americans just wanted more leisure and did not want more goods and services, higher productivity could lead to all of these benefits without increasing our production of goods and services.

In practice there is a real quarrel. There is no evidence that most Americans would want to use higher productivity to have more leisure and would not use their potentially higher income to buy more goods and services. As has previously been shown, there is an intimate connection between output growth and productivity growth. If output is not growing there is little need to build new plants or develop new processes. Countries where output is growing rapidly will automatically have higher rates of productivity growth.

While environmentalism is not commonly seen as an income distribution problem, it is closely linked with changes in the distribution of income. If you look at the countries that are interested in environmentalism, or at the individuals who support environmentalism within each country, one is struck by the extent to which environmentalism is an interest of the upper middle class. Poor countries and poor individuals simply aren't interested. If you reflect upon this phenomenon, it is not surprising. As our incomes rise, each of us shifts our focus of demand for more goods and services. Initially, we are only interested in physiological survival. Food constitutes our main demand. As we grow even wealthier, our demands shift toward roomier housing and higher quality food. At still higher income levels, demands rise for services. To get service we start to eat out in restaurants more and at home less. To avoid the drudgery of household work we mechanize household operations, and wives go off to find more interesting work outside of the household.

Suppose now that a family has reached an economy level where they can afford good food, fine housing, vacations, consumer durables, and all of those goods and services that represent the American dream. What is there left that can mar their economic happiness? Up to this point each of them can individually buy a rising real standard of living. But now they run into environmental pollution. If the air is dirty or noisy, the water polluted, and the land despoiled, there is a roadblock in their way to a higher real standard of living. They cannot achieve a higher real standard of living unless something can be done about environmental conditions. Environmentalism is a demand for more goods and services (clean air, water, and so forth) that does not differ from other consumption demands except that it can only be achieved collectively. In any geographic region, we either all breathe clean air or none of us breathes clean air.

From this perspective, environmentalism is a natural product of a rising real standard of living. We have simply reached the point where, for many Americans, the next item on their acquisitive agenda is a cleaner environment. If they can achieve it, it will make all of the other goods and services (boats, summer homes, and so forth) more enjoyable.

Environmentalism is not ethical values pitted against economic values. It is thoroughly economic. It is simply a case where a particular segment of the income distribution wants some economic goods and services (a clean environment) that cannot be achieved without collective action. Therefore, they have to persuade the rest of society that it is important to have a clean environment and impose rules and regulations that force others to produce a clean environment.

If you own a fine house and your neighbor dumps his garbage over the fence, you would call the police. If your neighbor burns his garbage (throws it up in the air) and it floats into your yard, current laws may not let you call the police, but you would want such laws. If you had the money to visit places of spectacular natural beauty, you would want places to visit. You want limits on development.

Environmentalism is the product of a distribution of income that has reached the point where many individuals find that a "clean" environment is important to their real standard of living. Just as it is not surprising that it is most in demand by the upper middle classes, so it is not surprising that it tends to be resisted by both the rich and the bottom half of the income distribution. Lower-income groups simply have not yet reached income levels where a cleaner environment is high on their list of demands, and it often threatens their income-earning opportunities. Very high-income groups can, to a great extent, buy their way out of the environmental problem, and they see environmentalism primarily as frustrating their efforts to earn even higher incomes.

A major part of the problem in the environmental area is that we are not used to thinking of a clean environment as a normal economic commodity. Environmental conditions have been excluded from our traditional measures of economic output for two reasons. Since they cannot be sold in private markets, it is difficult to determine exactly what they are worth. And, in the past, they may have had a zero price. If the water is clean, no one would be willing to pay for clean water—they already have it free. But neither of these reasons alters the fact that clean water is an economic good just as much as the private boat that sails upon it. Given the relative supplies and demands for a clean environment, environmental goods now have a positive price. They are a part of economic growth. They have not yet been included in our measures of GNP, except on an experimental basis, but this reflects measurement problems in calculating the GNP and not the economic merit of including them.

While it is possible to expand the meaning of the word "environmental" to the point where it includes everything (income, housing, and so forth), environmentalism ceases to have any meaning. It simply becomes another word for social problems. As a result, it makes sense to limit environmentalism to four major concerns: the pollution of air, land, and water; the exhaustion of nonrenewable natural resources; wilderness and species preservation; and the health and safety factors in industrial production. There are three questions that need to be addressed to each of these concerns. (1) What are the interrelationships between economic growth and the quality of the environment? (2) Should public policies seek to limit economic growth to improve environmental conditions? (3) If public policies are used to limit economic growth, what will happen to the distribution of income?

It is in the latter area that our basic zero-sum problem emerges. Since a clean environment is evaluated differently by different income classes, the comparison of costs and benefits will also differ markedly. Different groups can look at exactly the same costs and exactly the same improvements in the quality of the environment and differ on whether the costs exceed the benefits. Since we have to share a common environment with a common set of costs, environment expenditures inevitably end up raising the real income of income classes who have a clean environment next on their acquisitive agenda, and they lower the real income of those who have to help pay for a clean environment but do not place a high value on it.

Pollution and Economic Growth

The basic problem is not "limits to growth." As we have already seen, growth is limited. From 1947 to 1978 output was growing at 3.6 percent per year, real per capita disposable personal income was growing at 2.3 percent per year, and productivity was growing at 2.6 percent per year.[1] Limits exist because of the nature of the universe in which we live (effort is required to produce goods and services), the nature of individual decisions (how hard do we want to work, how many children do we want to have), and social institutions (taxes, regulations). Quite low limits now exist. The only question is whether we want to take deliberate actions to lower the limits even further.

Pollution controls are often opposed on the grounds that they lower economic growth and will reduce our real standard of living. This simply isn't true. Pollution controls only lower our standard of living if the costs of the controls are greater than the benefits of a clean environment. An efficient set of controls would raise the real growth rate. Using our present measures of economic output, while the costs of cleaning it up do appear, the benefits of a clean environment do not appear. But this is a problem produced by inadequate statistics and not a basic characteristic of pollution controls.

It is not surprising, however, that there is an argument as to whether these controls raise or lower our real standard of living. If the benefits have a high value to one group—the upper middle class—and have a much lower benefit to other socioeconomic groups (yet everyone has to share in the costs), different groups will see the desirability of the programs differently. The output of the program isn't equally valuable to everyone.

In principle, the problem is the same as that involved in national defense. Everyone has to pay, yet everyone does not put an equal value on having another one thousand missiles. For some taxpayers, national defense expenditures are our best buy, and for others they are our worst buy. It all depends upon your preferences for more missiles. Similarly the benefits of pollution controls depend upon the value you place on a clean environment.

There is a way, however, that each of us should think about the problem of how much "clean environment" to buy. Imagine that someone could sell you an invisible, completely comfortable face-mask that would guarantee you clean air. How much would you be willing to pay for such a device? Whatever you would be willing to pay is what economists call the *shadow price* of clean air. If we added up the amounts that each of us would be willing to pay for such a mask, we would have society's shadow price for clean air. Such a facemask cannot be purchased, but any pollution control program that can give us clean air for less than this price is a program that is raising our real standard of living. What we get in terms of benefits is greater than what we must sacrifice in terms of costs.

The *gedanken experiment* also tells us how environmental costs should be allocated. Revenue should be raised based on the amount that each of us would be willing to pay for our clean-air facemask. Those who place a high value on a clean environment would pay a great deal; those who place a low value on a clean environment would pay less. While we cannot actually perform our gedanken experiment, we should keep it in mind as we think of how the costs of environmental expenditures should be financed.

While there is no shadow price that is beyond controversy, the basic problem in our national debate about pollution controls is that neither side is really willing to sit down and place a value on a clean environment and then do the necessary calculations to see whether it can be had for less than this price. Until we do this, no one can say whether pollution controls accelerate or decelerate real economic growth. But it is important to remember that in principle, there is no conflict between pollution controls and economic growth.

Even output, as it is conventionally measured, may not slow down. This depends upon whether we finance our pollution controls by reducing other forms of consumption or by reducing other forms of investment. If we cut investment to make room for pollution expenditures, conventional output will grow less rapidly since the capital stock will grow more slowly. If we cut other consumption expenditures to make room for pollution expenditures, there is no reason to believe that even the growth of conventional output will fall. Pollution control devices are counted as part of our economic output, and we will simply have more such pollution devices and fewer other goods. Our total consumption will not be down.

But what about the reverse problem? Pollution controls may not adversely affect economic growth, but does economic growth adversely affect pollution? The obvious answer is "yes." Reductions in the level of economic activity below what they would otherwise be would clearly reduce the level of pollution below what it otherwise would be. The real question is whether ZEG is a first-best solution or an nth-best solution.

Slowing the entire economy to stop pollution is roughly equivalent to using an atomic bomb to swat a fly. Pollution would go down, but at enormous costs, since nonpolluting activities would be slowed along with polluting ones. Advocates of ZEG often try to squirm out of this problem by saying that they really aren't for ZEG everywhere–just in polluting activities. But what are polluting activities? It is not at all clear. Educational institutions do not look dirty, but they are large consumers of construction materials. Hospitals are prodigious users of polluting goods of all kinds. Direct pollution may be easy to identify, but indirect pollution is not. Each one of us is responsible for our part of the pollution caused by electrical power generation.

Even if we knew who was a polluter (we don't) and had selective controls to limit their expenditures (which we don't), what would we do? Would we place limits on their growth? If we did, firms would have no incentive to learn how to produce what they are now producing with less pollution. We would have locked society into its current pattern of pollution.

The preferred economic solution is a system of *effluent* charges where taxes are used to raise the price of polluting goods and services to a level consistent with the shadow price we place on a clean environment. If a one-dollar pad of paper generates twenty cents worth of pollution, we place a 20 percent tax on paper. This gives each of us an incentive to cut our consumption of polluting goods (paper now costs $1.20 and not $1.00), provides funds for cleaning up the environment or for more R&D on better pollution controls (twenty cents per pad of paper sold), and allocates costs to those that use polluting goods and services (the real standard of living is down $0.20 per pad of paper purchased).

Effluent charges are often resisted on the grounds that they let the rich buy the right to pollute. This charge is correct, incorrect, and irrelevant all at the same time. It is correct in the technical sense that anyone has the right to buy products that cause pollution. But it is a right that will make them poorer. It is incorrect in that the price will be high enough to discourage purchases and provide funds for public efforts to clean up the environment. The environment will end up cleaner. It is irrelevant because the extra amount that the rich will have to pay is larger than the value we place on a clean environment. If they choose to buy the right to pollute, they are transferring real income to the rest of us.

But whatever technique is used to reduce pollution, it is important to understand that the consumer is going to pay. Ultimately, firms pass along the costs of all inputs to their consumers. If they must pay effluent charges, they will raise the price of their goods to cover these charges or the costs of the facilities necessary to avoid these charges. If they are forced to stop polluting by rules and regulations, they will do the same. The costs of compliance, whatever they are, will ultimately appear in the price of the product. This is not bad but good. If the production of paper causes pollution, we are only going to use less paper if paper is more expensive.

The problem of industrial health and safety is identical to that of pollution; the only difference is that workers rather than consumers or neighbors are subject to damage. As before, the basic problem is one of measurement. Greater health and safety costs money, but we have not traditionally counted the benefits generated as output. No one doubts that this is difficult to do, but there also is no doubt that greater health and safety are desired by each one of us when we are personally involved. As with pollution efforts, an increase in health and safety will raise the price of those goods that are dangerous to produce. This will encourage us to use fewer dangerous goods and to shift to less dangerous alternatives. If done properly, the net result should be an increase in real economic growth.

Nonrenewable Resources and Economic Growth

While there is no direct conflict between pollution and economic growth, nonrenewable resources would seem to present a different problem. Theoretically, resource exhaustion could require a larger and larger fraction of our productive effort to produce a given quantum of raw materials as we are forced to retreat to lower-grade ores and less productive energy sources. Whether in fact nonrenewable resources do or do not act as a brake on economic growth depends upon a number of factors.

God could undoubtedly tell us the number of tons of each nonrenewable resource available in the planet Earth. It is a finite (but large) number; but it is also an irrelevant number. From the point of view of the economy, nonrenewable natural resources are actually growing because of economic progress in finding new ore bodies, extracting low-concentration ores, recycling used materials, and developing renewable substitutes (optic fibers for copper wires). Since with the exception of energy nothing disappears, we cannot physically use up any nonrenewable resources. They are always here. The total tonnage neither rises nor falls with use. The only question is whether we can use it economically.

Usable, nonrenewable resources supplies are expanding or contracting depending upon what is happening to relative prices. If prices are falling, resources are becoming more plentiful; if prices are rising, resources are becoming less plentiful. Excluding energy, the price of crude raw materials has fallen relative to finished goods from 1947 to 1978. In terms of the hours of work necessary to buy them, they have become much cheaper. While raw material prices were rising 2.6-fold from 1947 to 1978, per capita disposable income was rising 5.7-fold.[2] Measured in terms of work effort, raw material prices have been more than cut in half. The same picture exists over the past decade. Individual years can be found where prices rise, but the long-term and short-term trend is downward.

Even if this were not true, there would be little to worry about. As nonrenewable resources become more expensive, we would use less of them, find that we could now economically use sources that were previously too expensive, and develop substitutes. Natural resources are not a box of chocolates that we munch through and then are surprised that the box is empty. Long before the usable supply of any natural resource is exhausted, it will be so expensive that we are using very little of it.

Worries about natural-resource exhaustion are hard to rationalize from the point of view of economics. Depending upon relative supplies and demands, some natural resources will be cheap (sand) while others (diamonds) will be expensive. As demands for natural resources rise relative to supplies, prices go up. As prices go up the material will be used in fewer and fewer applications (copper). To some extent, other materials (aluminum) will be substituted, and to some extent products will simply become more expensive and less abundant (copper pots).

To argue that there is a natural resource problem, one must argue that for some reasons the market is selling raw materials too cheaply now, relative to future supplies and demands. But why should it do so? Those who buy and sell raw materials can make, and have every interest in making, the same calculations made by those who worry about resource exhaustion. If natural resources are going to be much more expensive tomorrow, one can make a great deal of money by waiting until tomorrow to sell.

Excluding energy for the moment, there seems to be no reason to believe that raw material prices will rise relative to other prices. They have not risen in the past, are not rising in the present, and there are no current signs that they will rise in the future. And even if they were to rise, this is one problem that markets are perfectly capable of handling. Slowly rising prices in response to long-run shifts in supplies and demands should place no undue economic strain on the system. Relative price changes are occurring all the time, and the fact that the good in question happens to be a nonrenewable natural resource presents no peculiar problems. If markets cannot handle such a problem, they cannot handle anything.

While there is every reason to believe that market expectations about the future are as good as anyone else's expectations about the future, this is not to say that the market is going to be right. What if it is wrong, and raw material prices are higher in the future than is now expected? As long as supplies are not suddenly cut off, prices will simply start rising at a later date and rise slightly faster than if more accurate predictions had been made. But there will still be adequate time to adjust, since any "natural" shortage of raw materials is going to be visible to everyone long before we have run out of any raw materials.

But what about energy? It has risen in relative price (2.6 times as fast as finished goods and almost 25 percent faster than per capita incomes), and it is lost in usage.[3] While the direct cause of the price is a man-made cartel, rather than Mother Nature, there are those who argue that the cartel has only speeded up what would have happened anyway. Won't it retard our future growth prospects?

In the case of energy, it is important to distinguish between the effects on (1) a rise in price, (2) the speed with which prices rise, and (3) the availability of supplies at the market price.[4] While economic markets are good at adjusting to slow, persistent changes, in relative prices they are not good at adjusting to sudden, large changes due to man-made scarcities and political events. There is no doubt that such events are disruptive to economic growth. As seen in chapter 2, it is a man-made problem that demands man-made solutions.

While it may not be the current case, it is worth thinking about what the impacts would be if energy prices were rising due to a natural scarcity. Here one has to ask himself whether there was anything to gain by limiting the consumption of nonrenewable energy sources below the level that would automatically occur with rising energy prices. Since we are going to have to shift to coal or renewable forms of energy (wind, solar, tidal, fusion) at some point, is there anything to be gained by delaying as long as possible the day when oil is too expensive to be used for heating?

If we limit growth to extend the period of cheap oil energy, the population gets cheaper consumption goods (less-expensive energy inputs are required), but fewer of these goods than it otherwise would have. Total consumption falls either way. Limiting oil consumption to achieve this result is only rational if we collectively know that leisure is more valuable than consumption goods even though each one of us would make the opposite decision, if allowed. But how could we come to such a decision in a democratic country? It is true that future consumption goods will become more expensive (require more work effort to produce the necessary energy), but this is true whether we do or do not limit the use of oil below the level called for by its price.

As long as you let energy prices reflect real current scarcities, there is no case for limiting the consumption of energy. There simply aren't any benefits. The basic problem still exists; society is running out of cheap oil, and the problem still has to be solved. The future real standard of living will fall, but for any individual worried about this problem there is a solution. Save and invest today's income to have a higher income tomorrow. Each of us, if we desire, can transfer consumption privileges from today to tomorrow.

Wilderness or Species Preservation

Wilderness or species preservation differs from the three previous concerns in that it has little or nothing to do with economic growth. The basic problem is one of consumption. How much of our potential output should we devote to protecting wildernesses or species? There is no technical economic answer to this question. It depends upon what constitutes your vision of a "good" society. To a great extent this will depend upon where you stand in the distribution of income.

When we allocate some of our resources to wildernesses, we are buying a particular type of consumption good. As long as we pay for it by reducing other forms of consumption, there is no reason to believe that it will affect real growth. Potential wilderness areas may include as yet undiscovered raw materials, but this does not affect the analysis. Opponents of wilderness areas often act as if these resources were being thrown away. This is simply silly. The natural resources in wilderness areas do not disappear. Any future generation that decides that natural resources are more important than wilderness areas is free to change the law. We may never want to use them, but they are available to be used. Although environmentalists do not like to think of them as such, wilderness areas are to some extent natural resource insurance policies and a technique for saving resources for the future.

There is no need for wilderness or species preservation to hinder economic growth, or for economic growth to prevent such preservation. On all dimensions, preservation represents a gift to the future. If future generations want wilderness areas and species, it is such a gift. If future generations want space and raw materials, it is such a gift. We can give a gift, but like any giver we cannot determine how the gift will be used after we are dead.

Economic Implications of Zero Growth

What are the consequences of deliberately limiting economic growth? Since the interest in ZEG springs from a desire to avoid depletion of nonrenewable resources and to reduce pollution, a ZEG economy is one where technical progress continues to occur. Gains can be made in the efficiency with which natural resources are extracted and used. New processes can be designed to reduce pollution. Industries rise and fall, but within a fixed total. The problems with a completely static economy are so numerous and so obvious that they hardly need analysis.

Fortunately or unfortunately, post–World War II American economic history is full of periods of zero or negative economic growth —1949, 1954, 1957–58, 1960–61, 1969–70 and 1974–75.[5] At this writing another seems eminent in 1979. Since history has provided us with repeated experiments in zero or negative growth, we need merely analyze these recessions to see what would happen. Given an increase in productivity of 2 percent per year, 2 percent fewer workers are needed each year to produce a constant level of output. In addition our labor force is growing by about 1 percent per year due to population growth and rising female participation rates. When these two effects are combined, zero economic growth leads to increases in unemployment of three percentage points per year. After a while, unemployment would be so high that workers would quit looking for work and participation rates would fall, leading to hidden unemployment rather than measured unemployment. But there is no way around the fact that ZEG implies rapidly rising unemployment under our current institutional arrangements.

As has been seen in the chapter on inflation, this unemployment burden would be shared unequally. This sharing would also become even more unequal as employers shifted to their most preferred workers in the normal process of turnover. Minorities, the young, the old, and women would carry the burden of a ZEG society.

It is also possible to analyze what happens to the distribution of income during a recession in order to see what would happen to the distribution of income if ZEG were imposed. The income gap between the twenty-five percentile of the population and the seventy-five percentile of the population (the interquartile range) for example, would rise by 0.2 percent per year for whites and 2.3 percent per year for blacks with no growth.[6] With a higher burden of unemployment, black family incomes would fall 6.5 percent per year relative to those of whites. Since the models that generate these results are derived from short recessions, it would be a mistake to multiply by one hundred to see what life would be like one hundred years from now, but these models do indicate the directions and magnitudes of the initial changes that would occur if ZEG existed. Without a doubt, a ZEG society would be a more unequal society than ours.

Parity among groups would become more difficult to achieve. In a ZEG world there is no way to employ more women without making more men unemployed. Which men are to be thrown out of work? With seniority patterns of hiring and firing, older workers would be protected in most cases, although those who are laid off would find it almost impossible to find reemployment. The young would find that the economy was not generating an expanding array of job opportunities and would have to wait for the old to retire or die.

If income were the only benefit flowing from work, and earnings were merely a necessary bribe to get individuals to suffer the discomforts of work, the problems created by zero economic growth would be easily solved. Some system of transfer payments could be devised that would sustain the incomes of those who became unemployed and encourage those who do work to work less and share the work more.

But jobs are more than just a source of money income. There are a whole host of consumption benefits that flow from jobs that have little to do with money income, such as friends, status, feelings of accomplishment, fame, and power. Many jobs in our economy would be worth fighting over even if they generated no income. To whom are these jobs to be allocated? This question exists in every society, but in a society with zero economic growth it is more intense. Society cannot generate new economic avenues to status, fame, fortune, and power through economic growth. For anyone to achieve any of these goals, someone else has to be displaced. Since we find it difficult to make a society work with a substantial zero-sum element, it is difficult to believe that we could make a society that was a pure zero-sum game work at all.

While some see a no-growth society as a happier, less competitive society, this is hardly an outcome that is foreordained. With few opportunities for advancement the economy might become less competitive. But the reverse is probably more likely. Where at least some of our energies were previously used to enlarge the economic pie, all of our energies can now be devoted to dividing a pie that has stopped growing. We know from other zero-sum areas of life that they can be some or our most competitive activities. Sporting events and gambling are zero-sum activities, yet they are marked by intense cutthroat competition. "Kill, kill, kill" is a not unknown sporting cheer.

A peaceful no-growth society could only be achieved if we could satiate wants. While it is logically possible to imagine a culture that could sustain satiated wants in the face of noticeably higher living standards in the rest of the world, there is no such culture now in existence. The demand for a rising real standard of living is virtually universal. The only exceptions are persons at the top of the economic totem pole.

Often a fallacious "impossibility" argument is advanced to imply that we have to limit economic growth. The argument usually starts with a question. How many tons of these or those nonrenewable resources would be needed if everyone in the world now had the consumption standards enjoyed by those in the United States? The answer is designed to be a very large mind-boggling number which convinces you that something has to be done to limit American consumption and that others can never achieve our standard of living.

What the question ignores is the fact that the rest of the world cannot have a U.S. standard of living until it has a U.S. standard of productivity. While consumption would go up by a large amount if this were true so would production. The world can only consume what it produces. When the rest of the world has our standard of living, they will be producing the extra resources necessary to have it. The relative prices of different products would undoubtedly change if this were true. We undoubtedly would be forced to shift away from an oil economy faster and do more recycling of materials if the rest of the world were growing more rapidly; but economic advances in the rest of the world do not depend upon cuts in our consumption.

While one can imagine changes in the structure of our economy that would prevent the rising inequalities that would exist in our current economy with ZEG, they are difficult to implement. The basic problem is rationing work when there are many more workers than jobs. Problems in work rationing are identical to those of any other rationing system. What is a fair distribution of work, and how can the rules producing this distribution be enforced? The U.S. work force is marked by a wide variance in the numbers of hours worked by different members of the labor force. Almost 6 percent of those employed work less than fifteen hours per week. At the other extreme, slightly over 7 percent of those employed work over sixty hours per week.[7] If one were simply to limit the total number of hours that anyone could work, only a small fraction of the work force would find themselves with lower earnings until the limit moved below forty hours per week. This, however, would put the entire earnings burden of ZEG on those who now work the most. Absolute limits on work could also encourage a rapid increase in the number of secondary family workers, with a consequent need to reduce the maximum hours of work even more than was originally indicated.

Another option is to cut everyone's hours of work proportionally. This has the questionable advantage of preserving the current distribution of earnings, but proportional cutbacks are impossible to administer except in short-run periods of time. Given a very rapid turnover in the labor force, workers would quickly start exaggerating the number of hours of work they were seeking in order to be assigned the number of hours of work that they actually wanted. The history of actual work patterns would rapidly fade out of existence. As a result, proportional cutbacks are not an administratively workable option over any extended period of time.

As a consequence, an absolute across-the-board limit on hours of work would seem to be the only long-run option. To prevent the induced increase in part-time workers, the limit would have to be set in terms of hours of work per lifetime rather than per week or per year. This would prevent families from evading the rationing system by increasing their number of workers in the paid labor force. Teenagers would not work to supplement their parents' income because to do so would reduce their own adult earning capacity.

The economic costs of absolute limits on hours of work depend upon one's estimates of the relative importance of talent versus the willingness to sacrifice hours of time. As long as we are simply talking about hours of time, there is no economic loss (other than extra training costs) when one person's time is substituted for another's. To the extent that scarce talent is involved, however, society is deliberately cutting itself off from the consumption of a unique resource. The more special the talent, the greater the cost.

The major enforcement problem would occur in the area of paid hours versus actual hours. There would be a strong incentive from both employees and employers to devote substantial amounts of time to unpaid "preparation for work" and then to pay very high rates for a few hours of actual paid time. This would allow employees to avoid restrictions on hours of work and enable employers to avoid the training costs of having more employees. As a result, there is no doubt that there would be severe enforcement problems. Work probably could be rationed, but there is no doubt that the end result would be a substantial increase in economic controls. Many individuals would have to be forced to do what they do not want to do.

If ZEG is taken seriously, it does not make much sense. "Small is beautiful" sounds beautiful, but it does not exist because it does not jibe with human nature. Man is an acquisitive animal whose wants cannot be satiated. This is not a matter of advertising and conditioning, but a basic fact of existence. To try to straightjacket human beings into "small is beautiful" is to impose enormous costs; yet these would yield only modest benefits in terms of less pollution and slower exhaustion of resources. Other techniques can achieve these results at a much lower cost. A society that cannot solve distributional questions in the current context would be required to solve distributional questions in a much more difficult context. Every increase in income, every promotion, and every advancement would require someone else to give up something he had.

The Distributional Conflicts

While environmentalism could easily lead to a higher-average real standard of living, it will not do so for everyone. For those who place a low value on a clean environment and must share in the costs, their real standard of living will fall. If a high-quality environment is purchased with ZEG, very sharp income reductions will be allocated to those who carry the greatest unemployment burdens. If a high-quality environment is purchased with effluent charges, income reductions will be allocated to those who now consume high proportions of polluting goods and services. If a high-quality environment is purchased with tax collections, income cuts will be allocated to those who pay the most taxes. In none of these three options is the payer necessarily the person who places the highest value on a clean environment.

The allocation of large income losses could be avoided if costs were allocated to those that place a high value on a clean environment, but this is difficult both economically and politically. Exactly who places what value on a clean environment and how should the necessary revenue be extracted from them? Even an approximate economic answer to this question is difficult but trivial compared with the political problem.

Environmentalists are not suggesting that they should pay for a clean environment because it is going to raise their real standard of living. They see pollution generated by someone else, and that someone else should pay the necessary cleanup bill. But "that someone else" thinks environmental cleanup costs more than it is worth. Yet we cannot raise the necessary revenue to clean up the environment unless we can agree on who should pay the bill.

Chapter 6

Spreading Rules and Regulations

ENVIRONMENTALISM naturally leads to rules and regulations. Automobile producers are to meet emission and fuel standards. Electrical generating plants are to control sulfur and fly-ash emissions. Industrial firms are to stop discharging wastes into streams. The list of environmental regulations is not endless but covers thousands of pages. And these regulations can be matched page for page by regulations in other areas. Individuals often worry about the growth of government expenditures, but government's greatest growth has undoubtedly been in the area of regulations.

While regulations have grown dramatically in the United States, we probably have fewer regulations than any other industrial country. But we impose regulations in an advisory legal system that makes what regulations we do have much more difficult to implement. Long-time delays are common as we fight our way slowly through the court system to find out what the regulations really require. Once a goal has been legislated into law, the fight has just begun. To a great extent the time delays and uncertainty that this process creates have a more adverse effect on the economy than the regulations themselves.

Occasionally regulations are overtly imposed to raise the income of some group (farmers) and to lower the income of some other group (consumers of food), but more often they are proposed on the grounds that they will accomplish some worthwhile social objective. Trucking regulations are defended by truckers since they raise the income of truckers, but the defense is based on the grounds that the regulations will provide cheaper or more reliable transportation to small communities. Steel producers would not have lobbied for steel import restrictions if these restrictions had not resulted in higher incomes for producers, but their rationale for doing so was "national defense." But whatever the overt objective, the implicit objective is always to alter the distribution of income and this is almost always the real reason for the existence of any regulation.

Because economic regulations are designed to raise the income of someone (and therefore lower the income of others), no one can say that a regulation is good or bad without a vision of what distribution of income should exist, and how this distribution ought to be created. In the abstract deregulation is a popular cause. Everyone is for it. In practice each of us opposes deregulation when it will lower our own income.

The Growth of Regulations

A large number of factors have contributed to the growth of economic regulations. To some extent they spring from our lack of other kinds of government involvement. In Japan, where industry is heavily dependent upon the Bank of Japan for its financing, government can issue marching orders to steel mills to stop air pollution without detailed regulations. The firms know that if they do not "voluntarily" perform the desired task they will have trouble with their "friendly" banker. In contrast, U.S. steel mills can only be persuaded to stop pollution with a host of complicated, cumbersome legal restrictions which must be legislated, drafted, enforced, and arbitrated in court over a substantial period of time. The same phenomenon is visible in Europe where governments own many industries (Volkswagen, Renault, British Steel), and firms need government's help in the capital market. When government owns or controls, it obviously does not need to write rules and regulations in the way we write rules and regulations.

Historically, there have been three great bursts of regulatory activity in the United States. The first occurred around the turn of the century. Antitrust laws were adopted to control man-made monopolies (oil, steel), and government regulations were invented to control natural monopolies (railroads, electrical power). In both cases the aim was to keep monopolists from extracting monopoly rents from the incomes of either consumers or other producers. The second burst of regulatory activity occurred during the 1930s. Various ineffectual schemes for curing the Great Depression by raising prices were adopted, but with the exception of agricultural price supports, the long-lasting regulations focused on constructing a legal environment that would make it easier for workers to form unions. The thirties essentially accepted the idea of large businesses but sought to limit their power with the countervailing power of large unions. The third burst of regulatory activity occurred during the late 1960s and early 1970s, and we are still in the process of digesting its effects. While the most recent burst has a number of facets, it essentially focuses on the problem of externalities and economic security.

In the case of externalities such as pollution, one individual can impose costs (dirty air) on another individual without having to pay compensation. The natural response of the second individual is to demand government regulations prohibiting the acts of the first individual, and this is exactly what has been happening.

Externalities have become important in our society for a number of reasons. As our society becomes more technologically advanced and more congested, one group's actions much more frequently impact another group's welfare—airport noise. But our technology has also exposed long-standing externalities that we previously did not recognize—the cancer danger of asbestos fibers. As we have seen, the rise in real incomes has also played a role. Environmental interests systematically depend upon our income. In the late 1960s the United States reached an income level where many people were rich enough to be concerned about a clean environment.

This economically induced shift in concerns was heightened by the fact that nature has some self-cleaning capacity. As long as the level of pollutants stays below this limit, Mother Nature cleans up the environment for us. If that limit is exceeded, however, the self-cleaning capacity often decreases. In the case of water, dissolved oxygen is the critical factor. When pollution rises above the level where self-cleaning is possible, the level of dissolved oxygen rapidly falls and water's self-cleaning capacity falls with it. The result is a situation where a small addition to the total amount of pollutants can lead to drastic increases in the amount of pollution. In the case of water pollution, most rivers were polluted well before the current interest in clean water; but in the case of automobile-induced air pollution, the effects are often recent. The city of Denver is a good example of a city that shifted from relatively clean air to heavily polluted air in a very short period of time. Air pollution is also of much more general concern. Water pollution can be avoided by staying away from polluted water, but it is much harder not to breathe than not to swim or boat.

We have already investigated the rising interest in income security. In most circumstances, regulations are seen as the best means for preserving or obtaining this security: foreign steel is to be kept out, set-aside programs are to raise the price of agricultural commodities, and entry is to be restricted into the trucking industry. But to provide economic security, rules and regulations must be issued. Farmers must be told what and how much to plant. Government ends up setting the price of steel depending upon its trigger-price regulations. To keep entry restricted in trucking detailed regulations must be written as to who can carry what. "Unroasted peanuts are not roasted peanuts."

The demands for protection have grown because, in a real sense, we have abandoned our belief in the virtues of a competitive, unplanned economy. Political speeches are still offered up to the totem of unplanned, competitive economies, but at the first sign of trouble everyone runs to the government looking for protection. The same steel executives who can give speeches about the virtues of competition ask for protection when competition arises. When deregulation of the airlines industry is proposed, the industry leads the opposition. Truckers may be in favor of competitive economics for others, but they want regulations for themselves. In some sense we have the worst of both worlds. We have a centrally regulated economy without the virtues of central planning. We won't admit what we are in fact doing, and thus every regulation is set up in isolation as if it were the only regulation in the system. We might gain in efficiency if we moved in either direction—toward more competition or toward more central planning.

At one time it was thought that all problems could be solved if the economy were only made competitive. It is this belief that lies behind both antitrust legislation and government regulatory agencies. The first is supposed to ensure that the basic conditions of competition exist, and the second is supposed to ensure that natural monopolies act as if competition did exist. For a number of reasons this vision has faded, but the regulations still exist and are now used to protect and raise incomes rather than to ensure competitive actions.

A competitive economy is without a doubt an economy filled with a great deal of potential opportunities, but many of these opportunities are bad. They result in income losses, risks, and uncertainties. We want opportunities for higher incomes (competition) but security for our present income (protective regulations). And as our consumption rises, the size of this desired, secure base has a habit of rising just as fast as our income. Unfortunately, we cannot have both. One man's security is another man's lack of opportunity. Thus we usually end up prescribing competition for others and security for ourselves. When we all do this, however, we end up with an economy full of regulations that prevent us from growing as fast as we should.

The attraction of the competitive ideal has faded for a number of other reasons as well. Individual firms and unions are so large that even if they are, in fact, competitive, the whole system does not have the attraction that it did when individuals were seen as the principal economic actors. General Motors versus Toyota may be real competition, but it is hardly a form of competition that wins the emotional support of the average man.

We also now realize that many of our problems would not be solved with more competition. To the extent that we have problems with externalities, competitive markets are no solution at all. A competitive firm will generate as much, or more, smoke than a noncompetitive one. There is also the suspicion that the virtues of small-scale competition ignore the problem of research, development, and economic progress over time. Without the research and development of oligopolistic firms, the economy might not grow as fast as it does. We pay a short-run price for some long-run gains.

Finally we now have historical experience demonstrating that the antitrust laws do not, in fact, produce competitive industries. At best the laws break one very large firm into two or three large firms after a very lengthy and costly legal battle, and the industry becomes slightly less oligopolistic. But slightly increasing the number of oligopolistic firms does not seem to make much difference in market behavior. The costs of enforcement are high and the benefits small. Antitrust laws have taken on a legal life of their own, but from the perspective of economics, they have little meaning or rationale. But with the intellectual heartbeat of antitrust dead, regulation remains as the only alternative. Instead of creating competition, we get into the business of trying to control oligopolistic behavior.

As new problems develop and conditions change, it is not surprising that new regulations have been written. This does not by itself, however, explain why the economy is becoming more regulated. If new regulations were matched by the abolition of old regulations, the economy would not become more regulated over time. Why aren't antitrust regulations abolished or substantially overhauled? Why aren't regulatory agencies abolished when conditions change? The failure of deregulation is central to understanding the regulatory process. To understand it, however, is to come right back to the problem of income protection.

Regulations persist since regulations affect incomes after they have been in place for any period of time. Someone's income is higher than it would be if the regulations did not exist. Deregulation, as a result, always poses economic losses for someone. Often the people who lose are not those who made the original income gains when the regulations were imposed. These are long dead or they long ago sold out the capital values that reflected the value of the regulations.

Workers have the same vested interest. They enter an industry expecting some wage, but that wage depends upon the existence of regulations. Regulations make the trucking industry profitable so that it can afford to pay its workers high wages. With deregulation, profits would fall and wages would be reduced. Over time, workers have acquired some of the benefits from charging higher prices to consumers, and over time they would lose some of those benefits if forced to give consumers lower prices. As a result, it is not surprising that both firms and employees resist deregulation strenuously. If the rest of us were to get cheaper products, they must earn less money. Politically, there may be more consumers than producers, but the per capita income losses to producers are much larger than the per capita income gains to consumers. Intensity overwhelms numbers, and the resistance to deregulation may be much stronger than the pressures for it.

Regulations are held in place by economic self-interest. In the case of airlines, this was true even when it was clear that profits would be higher after airline deregulations than before them. Total profits may be up, but some individual airlines will undoubtedly find themselves losing out in the new environment. Their profits will eventually fall. In addition, the new world is a world of uncertainty, and most are willing to trade some reduction in income for an increase in income security. In this, airlines are no different than the rest of us.

While it is relatively easy to chart the conflicting pressures that have led us into the current regulatory morass, it is much more difficult to chart a way out of the morass. What is the appropriate role for regulation? What goals can it not achieve? What goals must be achieved by other means? Before this task can be undertaken, it is necessary to clear the decks of some old intellectual baggage. Most of the debate surrounding regulation is encrusted in a set of issues and positions not relevant to the real problem. I have organized the cobweb-sweeping process under eight fundamental propositions about rules and regulations.

PROPOSITION I: ALL ECONOMIES ARE SETS OF RULES AND REGULATIONS—THERE IS NO SUCH THING AS THE UNREGULATED ECONOMY.

Often discussions of government regulations are posed in the form of a debate between the virtues of regulated versus unregulated economies. While debates are good clean fun, there is nothing to be debated if the issue is cast in this format. There simply is no such thing as the unregulated economy. All economies are sets of rules and regulations. Civilization is, in fact, an agreed upon set of rules of behavior. An economy without rules would be an economy in a state of anarchy where voluntary exchange was impossible. Superior force would be the sole means for conducting economic transactions. Everyone would be clubbing everyone else.

All market economies depend upon a set of rules and regulations that define how property rights are acquired and the conditions under which these property rights can be exchanged. One can have, and we have had, a market economy with or without the right to have slaves. Under current law one cannot even sell oneself into slavery. The elimination of slavery does not make the economy any more or less a market economy. It simply changes the domain over which individuals can enter into market transactions. In one case the market can deal in human bodies; in the other it cannot.

Before a market can be organized, the government must establish a set of rules and regulations specifying property rights. Without these regulations there is no theft (the illegal seizure of property rights), and without theft there is no room for a market. No one owns anything that can be exchanged for any other ownership claim. Therefore, the whole issue of property rights and transfer mechanisms must logically be considered prior to any debate about the merits or demerits of the market. Without government regulations there are no property rights, and without property rights there is no free market.

While property rights of long historical lineage often seem intuitively obvious, they are not. If my neighbor throws his garbage on my lawn, I have the right to call the police to stop him and the right to seek damages. If my neighbor throws his garbage in the air (burns it), I typically do not have the right to call the police and collect damages. I have property rights to land but no property rights to air. Yet clean air is probably more vital to my existence than clean space.

Each of us could give a good historical explanation as to why air property rights have not been developed. In a rural environment, air pollution is so unimportant that the ownership of air rights is simply not important enough to worry about. Clean air has a zero market price. But as society becomes more industrialized and heavily populated, clean air starts to have a positive market value, and its ownership becomes a real issue of concern. A similar change can be seen in the Law of the Sea Conference. It meets to define property rights to the ocean floor. But this issue only becomes worth discussing when we have the technology to mine the ocean floor. We do not debate the ownership of the planet Pluto since no one has the ability to appropriate it.

Technology may also lead to different specifications of property rights. Air moves around and is not easily appropriable in the way that land is appropriable. Do I own the air over my space or do I own some set of molecules? What happens if my air wanders? Air property rights have to be collective property rights rather than individual property rights. If they are collectively owned, then a set of rules and regulations for their collective use must be written. If they are individually owned, then a set of rules and regulations setting out the conditions of individual ownership must be written. But in either case, there are going to be numerous rules and regulations. The issue is one of writing a set of rules and regulations, which society can live by, in an area that has not previously been important enough to merit a set of rules and regulations setting out collective or individual property rights. This is not the place to argue whether the current clean-air rules and regulations are the right rules and regulations. It is the place, however, to point out that no matter how the issue is decided, there are going to be more rules and regulations than there were in the past.

Without the rules and regulations, the strong will simply seize what they want from the weak and then the strong will have to invest real resources in defending what they have just seized. Property rights exist to establish a just (or at least a widely agreed upon) set of rules for acquiring and exchanging property and to then reduce the costs of production and consumption by agreeing not to seize each other's property.

Thus the question of property rights is central to any economy, regardless of its degree of allegiance to market concepts. Property rights are, however, nothing but a set of rules and regulations. A private property economy is by definition a regulated economy. If the state owned everything and there were no private property, there would be no need for government regulations spelling out the nuances of property rights. There would need be only one regulation. All property belongs to the state, and it can do with it what it wishes. The real question and the real debate revolves not around the virtues of the regulated versus the unregulated economy, but around the question of what constitutes a good set of regulations.

PROPOSITION II: THERE ARE MANY SILLY GOVERNMENT REGULATIONS.

Gleefully finding silly government regulations has almost reached the status of a national parlor game. And nowhere is it easier to play the game than in the domain of OSHA—the Occupational Safety and Health Administration. Whenever silliness arises, it is well to ask why. It could arise because we are chasing after silly ends, or it could arise because we are using inappropriate means to achieve perfectly respectable ends. In the case of OSHA, the latter is clearly true. No one questions the virtues or the seriousness of reducing industrial deaths and injuries. The question is one of means.

Basically the problem is not one of stupid bureaucrats but one of trying to write universal regulations in an area where it is impossible. A regulation that makes sense in one context may not in another. Take the problem of providing toilet facilities for farm workers. A regulation that may be eminently sensible in a densely populated truck-farming area (a toilet every forty acres) with hundreds of farm laborers may not make sense on a Montana ranch where it is miles to the nearest person, and where there are hundreds of thousands of empty acres that seldom, if ever, see an agricultural worker. Yet for a set of regulations to be sensible in every section of a country as large as the United States, it would have to be so lengthy that it would be equally silly. Suppose that someone were to report that it took the government ten thousand pages of regulations to spell out the appropriate toilet facilities for every conceivable condition. Each of those regulations could be sensible, yet the aggregate effect is nonsense. The problem is using an inappropriate means to achieve a respectable objective.

It is also well to remind ourselves that silliness is not limited to public actions or to other individuals. Every example of stupid government action could be matched by a private example. The Edsel, for example, has entered our language as the paradigm example of a stupid action that wasted millions of dollars worth of resources. Boston's John Hancock building, with its falling panes of glass, was a fiasco from the day it was built. What would have happened if some government bureaucrat had built it? The U.S. steel industry misinvested millions on open-hearth furnaces when it should have been building oxygen furnaces. Most of us would have to admit, at least to ourselves, that we have made stupid mistakes in our own budgets.

We tend to excuse stupid private actions on the ground that the agents making the mistakes are wasting their own resources and that therefore their mistakes are their business. To some extent this is true, although it does not change the fact that all decision makers make mistakes. But to a very substantial extent, it is also not true. Managers of large corporations are not making mistakes with their own money. Both they and the government bureaucrats are playing the economic game with someone else's money. Whenever private or public managers make mistakes, they are going to be wasting someone else's money. And in general, it is much easier for a voter to replace a poor public manager than it is for a shareholder to replace a poor private manager.

Recounting examples of silly stupid actions is good clean fun, but in the end it doesn't prove anything about the virtues of regulation. No one doubts that there are both sensible and senseless regulations. The problem is to maximize the proportion of regulations that fall into the sensible category.

PROPOSITION III: THERE ARE MANY AREAS THAT SHOULD HAVE FEWER REGULATIONS.

What holds the regulations in place? The answer is simple—income security. Any long-standing set of regulations ends up raising the income of someone. And usually this someone includes some capitalists, some workers, and some consumers. When regulations are repealed, those individuals stand to suffer income losses. Other individuals stand to make gains, but the principle of territorial imperative applies in such cases. People fight harder to protect what they have than they fight to get something they do not have and have never had. Often this is compounded by the fact that the income gainers are widely diffuse while the income losers are highly concentrated. One fights much harder to protect a large sum than one fights to get a small sum. In addition, those who have been receiving help are usually well aware of this fact and have thoroughly entrenched themselves to repel attacks on their privileges long before anyone thinks of mounting an attack.

Consider the simplest case of a taxi medallion. Suppose these medallions (a permit giving one the right to operate a taxi) sell for $20,000. What gives these permits their value is that some agency holds the supply of taxis below the competitive level, and this leads to high profits. If medallions were simply issued to every potential taxi that met the required safety standards, medallions would have no value. They have a value because there is a monopoly supplier of medallions (the city) that is not issuing more medallions (is holding the number of taxis below the competitive level) and thereby is generating extra profits.

Suppose the city were now to deregulate taxis and shift to a free entry system. The value of the medallions would go to zero. Obviously, if you have purchased the right to operate a taxi for $20,000 you do not want someone destroying your $20,000 asset. True, there are more passengers than there are taxi operators. But each of them would only receive a small reduction in fares if the system were deregulated. The customer does not have enough financial interest to spend the time and money fighting for deregulation, but the provider has a very substantial interest in spending the time and money fighting for regulation.

Or take the more important case of general transportation. Transportation is probably the best example of an area that should be deregulated from the point of view of economic efficiency. The Interstate Commerce Commission was set up in 1887 when the railroads were a genuine natural monopoly. Not only were they monopolies, but they were run by individuals who believed in extracting their full measure of monopoly rents. Regulations and regulatory agencies were necessary to protect consumers and other producers.

But as time passed and we invented or perfected planes, autos, trucks, pipelines, and a host of alternative transportation systems, an industry that was at one time a natural monopoly has become one that could potentially be one of our most competitive. Instead of gradually abolishing the regulations that made sense in 1887, we gradually expanded them to include these new forms of transportation. Regulations which at one time had been used to hold prices below the monopoly prices which would have been charged in an unregulated market became regulations that were used to hold prices above the competitive prices that would have been charged in an unregulated market.

As a result, the president of the American Trucking Association defends regulations in an interview in *Fortune* magazine.[1] A similar reaction occurred in the deregulation of airlines. Who objected in an op-ed article in the *New York Times*? Not some fool who wanted to regulate everything for the sake of regulating everything, but the president of American Airlines.

Conventional wisdom maintains that rules and regulations have been used to stop railroads from effectively competing with these other modes of transportation, and that the impact of the system has basically been a transfer of income from the railroad industry to these other forms of transportation, and to trucks in particular. The regulations have certainly hurt railroads and helped trucks, but not in the way it is commonly envisioned. Railroads have been hurt not by regulations that stop them from competing with trucks, but by regulations that stop them from dropping unprofitable branch lines and services. Railroads are asking for the right to drop these activities, but they are not asking for general deregulation. They also like monopoly pricing under the protection of a regulatory agency.

Trucks and railroads have characteristics that make them optimal for carrying very different types of commodities. Railroads are optimal for bulk commodities (coal, grain, lumber, and so forth) that must be hauled over long distances. Trucks are optimal for packaged goods where door-to-door delivery and greater speed is important. When goods have some of the characteristics of both, we simply use truck-trains that take advantage of both sets of characteristics. The big gains going to truckers come not from protection from railroads, but from protection from farm trucks. There are millions of farm trucks that are only used part of the year in farming and could be used in commodity transportation if it were not for regulations preventing their use. If farm trucks were used in this way, interstate trucking firms would be forced to charge lower prices and live with lower incomes. Since this extra income has over time become divided between the owners of trucking firms and the Teamsters driving the trucks, both labor and capital have a vested interest in fighting deregulation.

If one asks why the regulations are used to stop railroads from dropping unprofitable services or from raising the prices on these services, there is a simple answer. Since railroads are the most efficient and cheapest mechanism for transporting bulk commodities, any community that loses its railroad service is going to find bulk commodities much more expensive. Communities fight to protect their real incomes, and this means opposition to deregulation. These particular consumers will suffer real income reductions when deregulation occurs, even though consumers in general will gain real income increases.

Regulations often cause cross subsidies where profits on one set of activities are used to finance losses on another set of activities. In this case, consumers in major population centers pay higher freight rates in order to lower freight rates in small population centers. Other examples of cross subsidies occur in the post office (rural areas and magazines are subsidized) and, if AT&T can be believed, in the telephone industry. Long-distance calls subsidize local calls. Business charges are used to cross subsidize residential charges.

While there may have been small income gains for some forms of transportation at the expense of other forms of transportation, the real income gains have come from setting up an industrial structure that makes it possible for all forms of transportation to raise prices above the competitive level and thereby extract some extra income from consumers—not from each other. Recognizing that any such system is always vulnerable to political attack, the transportation industry has for generations been active politically. Through campaign contributions and past favors, it has established a deeply entrenched position in both political parties that would require a lot of political capital to overturn.

On the other side, transportation charges are small enough and hidden in other prices so that no consumer thinks he stands to make a large gain in real incomes with deregulation. The aggregate gains spread across 220 million consumers may be very large, but no single one of these 220 million consumers may have enough of a financial interest to make it worth the time and effort to fight for deregulation.

As a result, the fact that there are many areas that should be deregulated does not lead to the conclusion that there are many areas that will be deregulated. General economic welfare may increase, but if our incomes were being threatened we also would be fighting deregulation. Efficiency is never a virtue when our incomes are being threatened.

PROPOSITION IV: IN THE UNITED STATES REGULATIONS ALMOST NEVER ARISE FROM IDEOLOGY—THEY ARISE FROM REAL PROBLEMS.

In many countries there are strong political parties committed to central planning, nationalization of basic industries, and regulatory control. Not in the United States. Here there are no strong political forces arguing for regulation for the sake of regulation. Regulations occur because the market fails to perform some task that the population wants performed or because our tolerance for failure has changed and we are no longer willing to tolerate long-standing failures. This principle can be seen in the infamous troika of EPA, OSHA, and ARISA (the Environmental Protection Agency, Occupational Safety and Health Act, and American Retirement Income Security Act).

The EPA arose because the automobile brought air quality standards to the top of the public health agenda in many parts of the country. Air quality levels were reaching intolerable levels. When school recesses have to be abandoned because the air is dangerous to health, as is the case in Los Angeles, it does not take a skilled political prognosticator to predict that there will be demands to do something. In this case the formation of the EPA was also helped by dramatic proposals to cut down the redwoods and flood the Grand Canyon. Air pollution was the catalyst that started the EPA, but once it was in place other long-smouldering issues such as water quality, endangered species, and wilderness areas bubbled to the top. Once some force is strong enough to break a political logjam, other issues will float downstream in the same current. The resistance to change has been broken and the advocates of clean air have a political interest in looking for allies.

Given some important issue around which the public can be mobilized, the nature of the system changes dramatically. Some new legislation and a new agency are put in place; but more importantly a professional cadre of individuals interested in, and dependent upon, environmental protection comes into being. Some of these people work for government, but many of them work for the lobbying groups (Sierra Club, Audubon Society, and so forth) that supported the original legislation. Others are volunteers who have made this activity into a central concern of their lives. The industries that produce air and water-pollution control equipment start to have an important stake in the regulations. High-sulfur coal producers in the eastern part of the United States help lobby to make those utilities burning low-sulfur western coal install stack scrubbers, since this will eliminate the competitive advantage of western coal. (More eastern coal will be burned and the incomes of the eastern coal industry will be higher.) The public's attention will inevitably turn to some other issue, but this professional cadre can keep part of the public mobilized for a long fight and can sound the alarm whenever a dramatic issue surfaces. When it is necessary to mobilize the public to save the Grand Canyon, the system does not go back to the status quo ante when the Canyon is saved.

OSHA arose in the context of dramatic industrial accidents (mine disasters in West Virginia, kepone in Virginia), and the development of new technologies that were able to determine the long-run effects of asbestos, cotton dust, and the industrial pollution. A wide variety of substances could be shown to cause cancer in animals. Individuals were no longer willing to tolerate the black lung and silicosis that their fathers had tolerated. Occupational health and safety standards were not getting worse, but they were not keeping pace with our expectations. Here again, once in place, the regulators are going to address issues other than the dramatic issues that led to the regulations in the first place.

The ARISA regulation arose for the same reason. A very substantial number of elderly American workers were not going to receive a pension check they had been counting on because the firm (and hence the pension fund) into which they had contributed had gone broke before they reached retirement age. Others found themselves fired right before retirement so that firms could avoid paying their pension. Any of us who had been similarly treated would feel equally aggrieved and would demand that the government do something. As a result, it is not surprising that the government did do something.

In each of these cases it is fair to ask whether the rules and regulations actually adopted were the best solutions to the problem; but this does not alter the fact that there was a problem that had to be solved in some manner. The moral of the story is that it is not very useful to be for or against regulations in the abstract. Something is going to be done to meet a clear and present need. The real trick is to find a set of regulations which solves the basic problem without creating a host of subsidiary problems that did not previously exist.

PROPOSITION V: THERE IS NO LEFT VERSUS RIGHT WHEN IT COMES TO THE VIRTUES OF REGULATION VERSUS DEREGULATION.

Sometimes the regulation debate is cast as if it is an issue of Left versus Right with the Left demanding regulation and the Right resisting it. In fact, the Left and the Right regularly switch sides depending on exactly what is being discussed. Consider the issue of drug regulation. What, if any, drugs should be banned? What, if any, drugs should be limited to prescriptions? We tend to think of drug regulations as if they date back to antiquity, but in fact they have been in place in the United States only since the late 1920s.

If we talk about deregulating marijuana, the issue is generally seen as a left-wing issue with the Left arguing for personal freedom while the Right argues for regulation. In fact, the issue is probably one that cuts across both conservative and liberal factions, with more than a few conservatives liking their occasional joint while puritanical liberals hold out for controls. If you take another drug such as laetrile, the issue is seen as right-wingers demanding the deregulation of the drug with left-wingers holding out for regulation. The Left holds that people should be protected from ineffective drugs, while the Right holds that personal freedom should reign. Here again, I suspect that in fact there are some of both on either side. And neither Left nor Right knows how to address our most dangerous and widely used drugs—tobacco and alcohol. Where do we enforce no smoking regulations? Should we subsidize tobacco farmers? What is the appropriate legal drinking age? Can blood tests be administered to automobile drivers? None of these questions can be solved in the context of Left versus Right.

Is the fifty-five-mile-per-hour speed limit a left-wing or a right-wing issue? In fact it seems to be a geographic issue. Are the rules and regulations protecting the U.S. steel industry a left-wing or a right-wing issue? In fact they are supported by everyone (Left, Right, and center) in steel-producing communities. Such questions could be multiplied many times. But the point is hopefully made.

There is no consistent left-wing or right-wing position on regulation versus deregulation. They are both for regulation under certain circumstances (often different circumstances, but sometimes the same), and they are both for deregulation under certain circumstances. Neither is consistent in its positions and neither has a firm grip on deregulatory virtue.

PROPOSITION VI: THERE IS NO SIMPLE CORRELATION BETWEEN THE DEGREE OF ECONOMIC SUCCESS AND THE DEGREE OF ECONOMIC REGULATION.

Despite the vocal complaints, the United States is by far the least regulated of all the industrialized countries. In other countries there typically is a large nationalized government sector (or sectors)—railroads and telephones everywhere, Volkswagen in Germany, steel and autos in England, Renault in France, and so forth). Elements of central planning are common. The Japanese have their MITI and the French have indicative planning. Government is often the principal source of capital investment funds and can control as it lends. The most extreme example of this is in Japan, where most investment funds directly or indirectly flow through government channels. Social welfare legislation and environmental protection are typically more advanced. Germany has its codetermination where union members must be represented on the board of directors.

Italy is famous for its complex rules and regulations. When it comes to the degree of regulation, the United States is a follower, not a leader.

Yet, as we have already seen, these economies are outperforming us and our own performance since the onset of the New Deal is better than our performance prior to the New Deal. Regulations can be deliberately used to promote economic growth, as in Japan, or they can be used to stop economic growth, as has been suggested by the ZEG forces. Either is possible.

PROPOSITION VII: REGULATIONS LEAD TO REGULATIONS.

Since individual economic actions occur in an integrated economy, the adoption of regulations in any one sector of this economy is apt to have effects in other parts. If you protect the steel industry and raise the price of steel, you are raising the costs of building cars in the United States and reducing the competitiveness of the U.S. car industry. Thus a regulation designed to protect one group is apt to hurt another group and lead to new regulations protecting the second group. If you protect steel, you are much more likely to have to protect autos. We have already examined the spreading wave of regulations in the energy area. Each new regulation forced us to yet add another regulation.

PROPOSITION VIII: LONG-RUN PRICE CONTROLS SHOULD BE AVOIDED.

Increasingly in our efforts to prevent redistributional changes from coming about, both liberals and conservatives are asking for price controls, protection, and subsidies for institutions—all in the name of protecting individuals. While price controls can be made to work in some economic systems, they cannot be made to work in our mixed economy with its openness to political pressures.

Basically, price controls should be avoided whenever society is addressing any long-range problem. Controls are only useful when society faces a temporary situation (such as a war or an embargo) that is going to cause a severe short-run disruption but will quickly end at some time in the future. This is a lesson that needs to be learned by both liberals and conservatives alike. Oil controls were imposed in 1957 to hold prices up and in 1973–74 to hold prices down. They were appropriate in neither case since the problem being addressed was long run in both cases.

First, let us consider the case where price controls are used to hold prices below the level that would occur in competitive markets. Such controls are almost always instituted to prevent some dramatic price increases that threaten to cause major reductions in the real incomes of some group. For some reason, there has been a sudden shift in supply and demand relationships, and people respond to the threat of real income reductions with demands for protection. Examples would include oil controls, natural gas regulations, and rent control. In the short run, price controls can mitigate the real income reductions, but at the expense of stopping someone else's income from rising.

If the situation causing the rapid escalation in price is temporary, a crop failure for example, then there may be a case for preventing the sharp temporary shift in the distribution of real income. Extra price incentives are not necessary to restore production in agriculture, and industrial wages might be disrupted if agricultural prices were allowed to rise.

If the problem is a long-run problem, however, a very different set of considerations prevails. Consider the regulations on natural gas. Because of OPEC and declining U.S. production, prices are not going to return to the old levels. Controls can be used to break one large real income shock into several smaller real income shocks over time, but they cannot restore the previous price level. When they are used to hold prices down for long periods of time, they cause perverse effects on both the supply and demand sides of the market. On the supply side, the incentive to look for new natural gas is reduced, and incentives are created to hoard old natural gas in expectation of higher prices in the future. The technique for alleviating the painful symptoms of the short-run shortages makes the long-run shortage worse and thereby increases the long-run pain. On the demand side of the market, the incentive to conserve is reduced. If other forms of energy (coal) are not regulated, there is in fact an incentive to use more natural gas since it is cheaper than competitive fuels. Once again this exacerbates the long-run problem.

Since the demands for natural gas exceed the supplies of natural gas at the controlled price, the system also requires rules and regulations for allocating the existing supplies. These rules and regulations usually end up allocating most of the available supplies to those users who have gotten natural gas in the past. But these are often not the people who can most efficiently use the now relatively scarcer supplies of gas. Industrial users (glass making) and home users who need clean sources of energy may not be able to get it while electrical power plants, who can use any form of energy, have it. As a result, the gas is not being used in the places where it could create the most value. The longer the price is held below its competitive level, the worse such inefficiencies become.

This situation also creates further distortions in the economic system. Since more natural gas is needed, gas suppliers sign contracts to bring in LNG (liquified natural gas) at prices far above the competitive price. They are allowed to pass imported prices along to the consumer, and thus the consumer ends up paying more for gas in the long run than if prices had been allowed to rise in the first place.

The same problems can be seen in rent control. Since New York City is the only major city that has had rent control for a long period of time, the problems are most visible there. First, those in controlled apartments have an incentive to hold onto the cheap controlled apartments even after family needs have changed. Large apartments are occupied by elderly couples or single individuals because they are cheaper than smaller apartments suitable to their needs. Thus, the effective supply of apartments is reduced and the upward price pressures on the noncontrolled part of the market is increased. On the supply side, landlords have neither the incentive nor the resources to maintain their buildings. Every price is appropriate to some quality level, and if the price controllers will not allow a landlord to raise his price to correspond to the current quality level, he can simply lower his quality level to correspond to the allowed price. Regulations can be written to try to stop this adjustment, but they are difficult or impossible to enforce. Other options also exist. If new buildings are not rent controlled, the appropriate strategy is for the landlord to let his building deteriorate (let the tenants gradually tear it down) and milk it for all of its cash flow during a period of deterioration. Once it has gone beyond the bounds of human habitation, it is torn down and a new uncontrolled building is erected. Or the building can be converted to condominiums. Renters are forced to become buyers at the market price or move out.

Since housing is subject to strong neighborhood externalities (the price of any one house depends upon the quality of the neighborhood in which it is located), any incentives to let housing quality fall can have large effects on others. Rents will fall in a neighborhood of deteriorating houses even if some of the houses are not deteriorating. And falling rents will mean that maintenance is not financially possible in the previously well-maintained buildings, and they will begin to deteriorate also. Long-run rent control is not the sole cause of New York City's housing problems, but it certainly has made these problems worse.

But there also is a question of equity involved. Price controls can only stop real income reductions by stopping real income increases for someone else. To hold down the price of natural gas or rents it is necessary to hold down someone's income. Let's suppose that we have decided to cushion the income shocks of market-priced natural gas or rents. This leads to the question of what is the fair way to raise the necessary revenue. Should we all have to pay through general taxation or is it fair to levy a tax on the owners of natural gas or apartments? They may be richer than the average consumer, but there also are many other even richer people who do not own natural gas or apartments. Why should they be excused from paying?

If price controls are effective in holding down prices, they end up creating even greater shortages and higher prices in the long run. Assets are frozen into efficient uses (LNG tankers). Efforts to legally evade the regulations generally lead to a set of perverse actions that host subsidiary problems (deteriorating neighborhoods). Th effective tax that is levied to pay for the income protection is ge erally an unfair tax in accordance with neither horizontal equ (the equal treatment of equals) nor vertical equity (the cor distribution of tax burdens between rich and poor).

Using price controls to hold prices above competitive le probably even more common than the reverse. Farm price su transportaton regulations, import restrictions, and minimum wages are only a few. In the long run, they create an equal set of problems.

Firms that benefit from controlled prices often do not have higher rates of return on capital than those that do not benefit. Initially profit rates are up, but this attracts more resources into the industry. Since price competition is prevented, these resources are typically used in various forms of nonprice competition–advertising, number of flights–or in excessive capital investments (too many oil wells, too many trucks). This use of assets is not completely useless, but it also is not the most efficient use of assets. The result is some net reduction in our rate of growth and real living standards.

If extra profits do emerge, these also lead only to a one-shot gain. The extra profits get capitalized into the price of the controlled assets, and any new purchasers receive a competitive rate of return on their investment. Taxi medallions have a value because they have succeeded in raising the profits of running a taxi cab, but the regulations do not help any new entrant into the taxi business since he must purchase a medallion as well as a taxi. When both costs are considered, taxis earn a competitive rate of return. But here again, no one who owns medallions wants their value reduced even if they do not raise long-run profits. The same is true in every other industry that enjoys protection.

The post office is another example. Since postal rates have been set at a level far above the cost of delivering much of the mail, regulations must be issued and enforced thereby stopping others from going into the first-class mail business. We are all familiar with the news story of the post office suing to stop some child from delivering local mail, but the purpose is to stop real competition. The result is a situation where the post office and postal workers have little or no incentive to cut mail delivery costs. As we have recently witnessed, this is true regardless of whether the post office is organized as a government department or as a profit-making corporation. Utilities, industrial mailing firms, and others would undoubtedly do to first-class rates what United Parcel and others have done to parcel post if given a chance. Undoubtedly there are routes in the United States that could not be competitively serviced with fifteen-cent first-class letters. But if low-price mail deliveries are a national goal, then we ought to finance this goal nationally and not levy a tax on those who should be getting cheap mail deliveries because they live in places where mail can be delivered cheaply.

In the short run, price controls are always seductive. They seem a direct way of achieving some objective, but they end up causing enormous long-run costs to achieve some very limited short-run objective. And if this objective is desired, there are many other ways (direct income subsidies) in which it can be achieved.

Generally we do not follow the cheaper, more direct route of direct income subsidies for a simple political reason. Voters will not put up with large direct subsidies, but large indirect subsidies can be hidden from the voter if rules and regulations are used. Agricultural subsidies are worth hundreds of thousands of dollars to some large farmers, but no Congress would ever pass a law simply giving large farmers several hundred thousand dollars apiece. The same Congress will, however, pass price support legislation that does exactly the same thing. Along with their greater efficiency, this is one of the reasons why direct subsidies should be used rather than indirect subsidies. If we would not support them overtly, probably we should not support them covertly. This applies to both industrial subsidies and human subsidies. If we want to raise the income of low-wage workers, a wage subsidy is to be preferred to a minimum wage. We then know what higher wages for low-income workers cost us and can evaluate whether it is worth the cost.

Where Do We Go from Here?

If we are to establish a competitive economy within a framework of international trade and international competition, it is time to recognize that the techniques of the nineteenth century are not applicable in getting ready for the twenty-first century. The late nineteenth and early twentieth centuries witnessed a two-pronged effort to create and maintain competitive capitalism. Antitrust laws were developed to break up man-made monopolies, and regulations were developed to make natural monopolies act as if they were competitive. While both of these approaches have had their problems, the time has come to recognize that the antitrust approach has been a failure. The costs it imposes far exceed any benefits it brings.

The futility and obsolescence of the antitrust laws can be seen from a number of vantage points. First, with the growth of international trade it is no longer possible to determine whether an effective monopoly exists by looking at local market shares. Regardless of the share of domestic production held by General Motors, General Motors is part of a competitive industry and must deal with strong Japanese and European competitors. In markets where international trade exists or could exist, national antitrust laws no longer make sense. If they do anything, they only serve to hinder U.S. competitors who must live by a code that their foreign competitors can ignore.

One could debate whether international antitrust laws would make sense, but this debate would be completely irrelevant from a practical perspective. In the absence of anything resembling world government, and in the presence of widely differing views on the usefulness of antitrust legislation, no enforceable, international antitrust laws are going to come into existence.

If competitive markets are desired, the appropriate policy should be to reduce barriers to free trade. Whatever good competitive effects the antitrust laws have had on the behavior of the U.S. steel industry, they are completely dominated by the bad competitive effects of the reference price system designed to keep foreign steel out of the United States. Whatever good competitive effects the antitrust laws may have had on the behavior of U.S. auto makers, they are small in comparison with the competitive pressures brought by Japanese and European automobile producers. If one measures the potential gains to be made by enforcing the antitrust laws, as opposed to reducing real barriers to international trade, it is clear that the large gains exist in the area of more international competition.

Second, as incomes rise it becomes less and less clear as to what is the relevant market to determine whether a firm has acquired a monopolistic position. Most goods we buy are not physiological necessities but luxuries that could be substituted by other goods to produce just as high a real standard of living. Rolls Royces and Volkswagens are both cars, but the two products are in no sense competitive. For those who buy an expensive car, the real trade-off may be with a swimming pool, a summer home, or a wide variety of other products. Rolls Royce may have a virtual monopoly position in the production of very expensive cars yet still not have a position it can exploit. If it prices its product too high, people will shift to different products.

As an illustration of the same problem at the other end of the price spectrum, consider the antitrust case in the breakfast cereals business. Let us assume that a few companies have established an oligopolistic position with respect to dry breakfast cereals and are charging more than would be charged in a competitive market. Since any individual consumer can, if she or he chooses, buy no-name brand corn flakes at a much lower price, the brand names must be yielding some psychic utility or brand name corn flakes would not be sold. Consumers may have been convinced of this psychic utility because of advertising, but so what? At the income level of most Americans, most wants have been determined by some explicit or implicit form of advertising. Psychological needs determine very few of our expenditure decisions. Individual consumers may be making silly decisions (buying products at prices higher than they need to pay), but it is hardly the appropriate role of government, much less the antitrust laws, to stop people from making silly decisions that do not affect anyone but themselves.

But let's suppose that the no-name brands did not exist. Since corn flakes are hardly a unique, patented, hard-to-produce product, the absence of no-name brand corn flakes could only mean that individuals are willing to pay for having brand name corn flakes. People are allowed to pay for the brand labels of clothing designers. Why stop brand labels here? If the brand premium gets too large, others can easily enter the no-name brand corn flakes market. But even if no-name brand corn flakes could not be produced, there are still a great deal of other breakfast alternatives (bacon and eggs, no breakfast). These other products make the market a competitive market even if there is no competition within the dry cereals business.

Third, monopoly rents are inherently limited in an economy full of large conglomerate firms. Since established market positions are usually easier to defend than to create, oligopolistic firms may be able to extract a small price premium from their customers, but this ability is inherently limited by the ability of other large firms to enter the market. Excessive rates of return attract competitors, and potential competitors have the ability to enter all those markets that are not natural monopolies. While the conglomerate movement may have lumped activities together that are not the most efficient sets of activities to be lumped together (to the same extent this is a product of the antitrust laws), it has also created a set of large firms that scan a wide range of products and markets to search for profitable investments. Firms with no actual or potential competitors are few and far between. As a result, this apparent monopolistic position is actually vulnerable from both the demand and supply side of the market. Potential customers have alternative uses for their incomes and potential competitors are almost always waiting in the wings if profits appear too high.

Fourth, it is not obvious that anything of economic value is accomplished even if an antitrust case is won by the government. Consider the current IBM case. Suppose the government were to win and IBM were to be broken into three or four large firms (an outcome that is highly unlikely given recent antitrust experience). What characteristics of the industry would change? By now we should have enough experience to know that a three- or four-firm oligopoly does not act noticeably different from a one-firm monopoly faced with potential competition (the Japanese) in its main business and actual competition where it is weak (in small computers).

If you look at other industries where antitrust laws have resulted in the creation of new competitors (oil and aluminum) or where they have stopped inefficient producers from being absorbed by competitors (steel and autos), it is hard to argue that these industries are more efficient or less competitive than the computer industry. IBM has driven other large firms out of the industry (GE and RCA) through being able to provide a better product. If the case were to succeed, the most likely winners would not be computer customers but foreign computer manufacturers. No one questions that IBM has a dominant market position. But this is not to say that it has been able to extract crippling monopoly rents from computer customers. In some ways the case reads like a government sign saying, "It does not pay to be too efficient." Yet in a larger context, this is certainly a slogan that we do not wish to issue if we are interested in long-run efficiency.

Fifth, the whole antitrust vision springs from a very narrow view of competition. Competition means price competition—nothing more and nothing less. Yet price is clearly only one of the many competitive weapons (advertising, product quality, and so forth) and in many areas not the most useful or used weapon. We have a vision in the backs of our minds that if we only create enough firms, firms will be driven to price competition and have to abandon other forms of competition.

There are several problems with this vision. Even if it were true, the required number of firms is so much larger than the number that would be created by an antitrust "win" that it has no relevance to antitrust legislation. But more fundamentally, it is not true. There are many industries with thousands of small-scale producers (real estate agents, lawyers, doctors, specialty shops) who do not compete based on price. Many customers would rather shop in elegant surroundings than buy at the lowest possible price. Shopping and the thrill of being enticed may be a major part of the enjoyment of buying goods and services. To look simply at the degree of price competition in the economy is to grossly underestimate the degree of real competition in the economy.

Somehow lurking in the backs of our minds is the puritan idea that if we could only strip away advertising, fancy surroundings, nonessential product characteristics, and the attractive salespersons, we would get back to true preferences that would create more enjoyment. Most of us think that we are clever enough to avoid being duped into doing anything that we do not really want to do, yet we think that we must act to protect someone else from being led astray. Why? Let me suggest that there is no reason. Nonprice forms of competition are just as useful and valid as price competition. When industries do not engage in price competition, there usually is a perfectly good reason (other than monopoly) as to why they do not. It simply isn't the most efficient way to compete. As a result, we are not going to restore price competition and puritan simplicity through the antitrust laws.

Given our modern economic environment, antitrust regulations should be stripped back to two basic propositions. The first would be a ban on predatory pricing. Large firms should not be allowed to drive small firms out of business by selectively lowering their prices in submarkets while they maintain high prices in other submarkets. The second proposition would be a ban on explicit or implicit cartels that share either markets or profits. Firms can grow by driving competitors out of business or by absorbing them, but they cannot agree not to compete with each other.

In an economy as complex as that of the United States, no one can with 100 percent certainty say exactly how competition and industrial structure would change if the antitrust laws were reduced to the fundamentals of prohibiting predatory pricing and cartel formation. Perhaps we would find that some unacceptable results would occur and that some additional regulations would be necessary. If so, the necessary regulations can be written when the abuses appear. This is an area that has become so complex and so little connected to economic goals that we need to start over and see what a modern economy needs in order to remain a competitive economy.

None of this is to deny that technological advances will from time to time require changes in the rules of industrial competition. With the development of microwaves and satellites, the long-distance transmission of messages may have changed from a natural monopoly to a potentially competitive industry. If so, rules and regulations governing the telephone business should be changed to reflect this development. Whatever should be done, however, the correct answer is not an antitrust case against AT&T. A regulated monopoly should be governed by regulatory procedures and not by antitrust procedures. If the goal is a competitive industry in long-distance transmission, an antitrust case is simply not the means for getting to this objective. Deregulation is best achieved by deregulation, not by a lengthy court case based on principles that have nothing to do with regulation or deregulation.

Basically, regulators have two sets of instruments. They can attempt to set regulations that influence the production of some good or service. These might be called *q* regulations since they are intended to affect quantities directly. Or, they can attempt to levy taxes or subsidies to encourage or discourage the production of some good or service. These might be called *p* regulations since they are intended to affect prices. In general, we have relied too much on *q*-type regulations and not enough on *p*-type regulations. With *p*-type regulations the regulator tries to take advantage of market incentives rather than attempting to fight market incentives as is the case with *q* regulations.

Effluent charges are *p* regulations that lead to less pollution. Insurance charges under the workman's compensation system can be raised to cover the real costs of industrial accidents, to allocate those costs to industries with poor health and safety records, and to give those industries a real incentive to improve their health and safety performance. Private pension funds can be federally insured in a system that reflects the real probabilities of default. Or individuals could be allowed to buy extra pension benefits through the social security system. Such charges have the advantage that they encourage each individual economic actor to find the best method for reducing pollutions and accidents or for assuring future pension rights.

The virtues of *p* regulations over *q* regulations are something that the economics profession has stressed for decades. There are undoubtedly some areas where such market solutions would be difficult to implement (cancer agents with long-time delays might be one), but they could solve the problems in many areas where regulations are not rampant. Yet despite our professed social belief in the value of markets, such solutions are resisted by almost everyone involved. Society has almost universally employed *q* regulations when it sought to achieve some objective. Why? Understanding and curing these preferences is at the heart of the issue. Why are we as a society so resistant to the idea of *p* regulations? Is it simply ignorance or is there something we wish to achieve that cannot be achieved with *p* regulations?

If one goes through the list of objections to *p* regulations, it is clear that they spring from a set of conflicting visions of the economy. As we have already seen, those who want to protect the environment or industrial workers often falsely believe that firms will simply pay rather than reduce pollution or accidents. *Q* regulations seem more certain to achieve the desired objective than *p* regulations. Given our recent history, the basis for this latter belief is difficult to fathom. Many *q* regulations have been written, but the objectives have not been achieved. Loopholes have been found (trucks and vans replace cars) or the regulations have been junked (compulsory seat belts) when they prove unworkable or intolerable. Writing a regulation does not guarantee that some objective will be accomplished. And by now we should have had enough practical experience to convince anyone that this is true.

P regulations are also opposed on the ground that they will simply be passed on to the consumer. This is both true and desirable and does not in any case distinguish them from *q* regulations. From the perspective of those wishing to reduce pollution or accidents, there is a final charge that *p* regulations are unfair, since they allow the rich to buy the right to pollute or cause accidents. As we have already seen, this is right, wrong, and irrelevant all at the same time.

From the perspective of those who are resisting pollution or safety regulations, there is usually just resistance without any participation in the debate about appropriate means. To get into the debate between *p* and *q* regulations would be to undercut the arguments against both *p* and *q* regulations. The resisters do not want regulations and they do not want to pay for something that they have always had free (the right to pollute). Blind resistance is only rational, however, if you believe that the programs can be defeated or avoided and frustrated at low cost.

Here again we have recently accumulated a massive amount of empirical experience. Whatever the original validity of the belief that the programs would go away or could easily be frustrated, the belief has been proven wrong by history. Regulations will be adopted since a majority wants to achieve the stated goals. These regulations can be frustrated so that the end results are not achieved, but very substantial costs will be imposed regardless of whether the regulations do or do not produce the desired results. And if the ends are not achieved, the result will not be a return to the status quo ante, but the adoption of a new and more stringent set of *q* regulations.

From the point of view of a resister, it is obviously much easier to live with a *p* regulation than with a host of *q* regulations. Yet it is very difficult for a resister to play a constructive role in debates about appropriate techniques for achieving a goal that he or she was seeking to frustrate. Whatever they recommend will always be seen as a vehicle for frustrating the desired objective rather than as a technique for achieving the desired objective with fewer undesirable side effects. The business community would have been better off to abandon a policy of total resistance and instead concentrate on influencing the means toward developing techniques with which they could more easily live. Getting to a set of *p* regulations after a set of *q* regulations has been adopted is going to be one of the major challenges of the 1980s.

Distribution Once Again

Since the ultimate aim of every competitive firm is to establish a monopoly position so that it can earn more than the competitive rate of return, rules and regulations are needed to ensure that not too many competitive firms succeed in achieving their objective. At the same time rules and regulations can be, and in many cases have become, the means whereby competitive firms and industries achieve their monopolistic goals. By stripping industrial rules and regulations down to the bare essentials, there is much more to be gained than to be lost. If we go too far and abuses emerge, they can be corrected. If new abuses do not appear, this is in fact a piece of evidence showing that we have not stripped away enough of the rules and regulations encrusting our economy.

But in the end the central problem reemerges. Whatever technique is chosen to reach different social objectives, someone's real income is going to go down. Rules and regulations are no exception to this dictum. The distribution transfers from one group of Americans to another group of Americans are not as overt and visible as they are in the area of direct income transfer payments, but they are undoubtedly larger. Without a vision of what constitutes an equitable distribution of income, it is not possible to say whether we have a good set of rules and regulations or how this set should be modified.

Chapter 7

Direct Redistributional Issues

ALL WESTERN GOVERNMENTS have become heavily involved in altering the distribution of market resources directly as well as indirectly. Ethically we have been committed for many centuries to the idea that a "good" society does not let families starve on the street. With this commitment comes the responsibility for providing a minimum family income if the family is, for whatever reason, unable to take care of itself. Exactly how this should be done and what level of minimum income should be provided is one of our most contentious issues, but it is unavoidable. Our ethics force the issue upon us, but it is also not at all clear that an industrial society with its delicate social and physical interactions could tolerate extreme deprivation. It would simply be too easy for those with nothing to lose and everything at stake to disrupt the rest of our society and economy.

With the collapse of the idea that market incomes are determined by impersonal forces outside of human control, direct redistributions have also extended far beyond that of simply establishing a minimum income floor to prevent extreme deprivation. In 1978 direct transfer payments accounted for $224 billion in annual spending.[1] Over 10 percent of our GNP was devoted to taking income from one private individual and giving it to another private individual.

If one looks at the consistency in the distribution of family income (see table 7–1), one might ask why direct redistributions are so contentious.[2] Those opposed to government income transfers have little reason to complain since the distribution of income has not been made more unequal. The rich are as rich as ever. Those in favor of government income redistribution have little reason to ask for more redistribution since $224 billion in annual expenditures have not succeeded in making the distribution of income more equal. If $224 billion does not solve the problem, why will more money solve the problem?

TABLE 7–1
Distribution of Family Income, 1947–77

	Shares (%)	
	1947	1977
Lowest Quintile	5.0	5.2
2nd Quintile	11.9	11.6
3rd Quintile	17.0	17.5
4th Quintile	23.1	24.2
Highest Quintile	43.0	41.5

SOURCE: U.S. Bureau of the Census, *Current Population Reports, Consumer Income 1977,* Series P–60, no. 118 (March 1979), p. 45.

Redistribution is a contentious issue and is going to become an even more contentious issue for a simple reason. The consistency in the distribution of family income has masked a market distribution of income (see table 7–2) that is slowly becoming more unequal. The bottom 60 percent of the population has been losing earnings, but its income has been held even with the rest of the population through the rapid rise in income transfer payments and labor force participation rates for women in these economic classes (see below).

In the future the trend toward a more unequal distribution of earnings is apt to show up in a more unequal distribution of family income. Labor force participation rates are now rising most rapidly for women who are married to men with high incomes. Although income transfer payments have stopped the economic gap between

TABLE 7–2
Distribution of Wage and Salary Earnings for Persons

	1948 (%)	1977 (%)
Lowest Quintile	2.6	1.7
2nd Quintile	8.1	7.7
3rd Quintile	16.6	16.1
4th Quintile	23.4	26.4
Highest Quintile	49.3	48.1

SOURCE: U.S. Bureau of the Census, *Current Population Reports, Consumer Income 1977,* Series P–60, no. 118 (March 1979), pp. 226, 227.

the rich and the poor from rising since World War II, they cannot continue to rise as fast as they have over the past two decades. With income transfer payments slowing down and working wives contributing to inequality rather than equality, the distribution of family income will start moving toward inequality in the 1980s and 1990s.

Given rising inequality, direct income redistributions are apt to become even more divisive than they now are. The zero-sum economic game is going to become harder to play since more direct income transfers are going to be demanded. These demands may not be met, but they will have to be faced.

To look solely at the tensions within the income transfer system, however, is to miss much of the direct redistributional problem. Governments impact the distribution of economic resources through two other direct channels. Whenever governments extract taxes they lower someone's income, but whenever that money is spent they also raise someone's income. Since governments do not spend their money on the same goods and services purchased by private individuals, a transfer of purchasing power from individuals to government yields a different set of demands, and in doing so a different distribution of market incomes. Demands go up for some skills and down for others.

Perhaps it is not surprising in a democracy, but each of these channels delivers most of its benefits to different economic classes. The poor have their income transfer payments and the rich have their tax loopholes, which allow them to pay less than their fair share of taxes. While the middle class looks down at the poor and up at the rich and often thinks that it pays for government without getting its share of the benefits, this belief is fallacious. Civilian government expenditures (roads, schools, parks) are heavily focused on the needs of the middle class, but even more important, government is a major provider of middle-class jobs. Without government there would be fewer middle-class jobs, fewer middle-class incomes, and fewer middle-class families.

This delicate political balance is being disrupted and it is not clear what will replace it. With inflation there is a demand to cut back on government expenditures. But any cutback will increase the economic pressures on the poor (fewer income transfer payments) and the middle class (fewer good jobs). With slow growth there are demands to further cut taxes on the rich to encourage savings and investment. But any cutback will necessitate increased taxes for the middle class. Pressures exist to dismantle the current troika of benefits, but what happens when the current system has broken down? With what do we replace it?

As large as the gap between the rich and the poor may be, the major demands for redistribution exist not on this dimension but between ethnic or sexual groups. Should government economic policies focus on eliminating differences in the economic outcome among such groups or should it focus on helping individuals whose economic performance is in some sense below society's norms of acceptability. This is a fundamental ideological question facing the United States and most other Western industrialized countries. Both our political and our economic traditions have historically focused attention on the individual. Individuals are awarded voting rights and individuals are to have an equal opportunity to achieve economic success. But ours is a society of groups, each demanding a larger fraction of the national pie. These demands have often been ignored in the past but can they be ignored in the future?

The Success or Failure of Current Redistributions

When we talk about the growth of government expenditures since World War II, we are talking about the growth of direct income transfers. While government purchases of goods and services were growing from 18.9 percent to 20.6 percent of the GNP from 1956 to 1978, income transfer payments were growing from 4.1 percent to 10.7 percent of the GNP.[3] Government purchases are up slightly (federal purchases are actually down from 10.9 to 7.3 percent of the GNP), but we are entering a period of falling purchases. School enrollments are beginning to fall, and over 40 percent of all state and local expenditures are for schools.

While the social welfare expenditures of the late 1960s and early 1970s are often described as a failure, they were, in fact, extremely successful. And nowhere is this more true than among the elderly. As has previously been examined in the chapter on inflation, the mean per capita income of the elderly now equals that of the rest of the population. If in-kind aid (medicare, food stamps) is considered, the elderly have a higher per capita income than the nonelderly. The percentage of the elderly living in poverty (14 percent) is slightly higher than that for the whole population (12 percent), but not much. If social security were to disappear for the elderly, their incomes would fall by 50 percent. Half of the income going to the elderly comes from government transfer payments.

Among the rest of the population, income transfer payments have been equally successful. The distribution of family income is a misleading indicator of success or failure for a number of reasons. First, it ignores 23 million single individuals who are not in families. Second, it does not take into account the fact that the average family has grown smaller since World War II. Birth rates are down, young people go off to set up their own households at an earlier age, and the elderly no longer live with their children. This creates more separate households with fewer earners per household.

A better indicator of real welfare gains is the distribution of per capita household income (see table 7–3).[4] Viewed from this perspective there has been a significant income gain for the lowest three quintiles of the income distribution. Where the top quintile had eleven times as much income as the bottom quintile in 1948, it had only seven times as much in 1977. This occurred in the face of rising inequality in earnings (the top quintile of earners had nineteen times as much as the bottom quintile in 1948 and twenty-eight times as much in 1977).[5]

TABLE 7–3
Distribution of Per Capita Household Income

	1948 (%)	1977 (%)
Lowest Quintile	4.1	5.6
2nd Quintile	10.5	11.7
3rd Quintile	16.0	18.1
4th Quintile	23.5	26.5
Highest Quintile	45.9	38.1

SOURCE: U.S. Bureau of the Census, *Current Population Reports, Consumer Income,* Series P–60, no. 117 (Dec. 1978), p. 22.

Without income transfer payments, the share of income going to the bottom quintile of households would have been more than cut in half during the post–World War II period. Government actions prevented this from happening and actually caused a substantial gain in the income position of the poor. Here again, making a correction for in-kind aid would make the gains even larger. While it is difficult to place a cash value to the recipient on many in-kind aid programs, some programs, such as food stamps, are fundable and just as valuable as cash. If you do not need the stamps you can sell them (or the food purchased with them) to someone else. In fiscal 1978 food stamps represented a $6 billion increase in the real income of low-income individuals. This represents a 25 percent increase in the income of the bottom quintile—a source of income that is not considered in official income statistics.

In addition, manpower training programs have kept the distribution of earnings from becoming even more unequal. In 1977 approximately $10 billion worth of wages and salaries were paid out under manpower training programs.[6] Almost all of this goes to workers in the bottom 40 percent of the work force, and it accounts for 18 percent of their earnings. Without manpower training programs the share of total earnings going to the bottom 40 percent of the work force would have fallen from 10.7 percent to 8.0 percent rather than to the 9.4 percent that actually occurred. The deterioration in the earnings power of the poorest workers was not halted, but it was significantly arrested.

While income transfer programs have kept the bottom quintile of the population from losing ground, working wives stopped the second and third quintiles from losing ground to the top two quintiles. It is among this income class that female participation rates are the highest, have risen the most, and are adding the most to family earnings.

This source of household income equality is probably already in the process of vanishing. Under the impact of female liberation and the general long-run trend toward female work, female participation rates are now rising most rapidly for wives of high-income husbands. Since participation rates have farthest to rise for wives of high-income husbands, high-income families stand to gain the most from future increases in female participation rates. If you believe in selective mating (men are married to women who make the same amount in a nondiscriminatory, equal participation world), then participation effects are apt to be magnified by larger earnings gains for wives of high-earnings husbands. At the moment, the earnings differences for wives of high-earnings husbands and wives of low-earnings husbands are not substantial. Working wives contribute to equality since their earnings are much more equally distributed than those of their husbands. In a nondiscriminating, equal participation world, female earnings are apt to be as unequal as those of men. The net result—a more unequal distribution of household income.

As a consequence, one of the two sources of constant family income shares is about to reverse itself and become a source of greater inequality. We are entering a period of rising inequality where conventional income transfer programs will be incapable of preserving the current degree of inequality. And as the second and third quintiles of the income distribution lose ground to the top two quintiles, they are unlikely to support the increases in income transfers that would be necessary to keep the first or poorest quintile from losing ground.

The bottom three quintiles will be losing ground to the top two quintiles of the population. In the past two decades we have coped with rising market inequality by increasing income transfers sharply and providing jobs for the wives of low-wage husbands. Neither solution is apt to be possible in the next two decades. How are we going to cope with rising inequality in household incomes? No one knows because we have never had a period of rising inequality since income data became available.

Economically there is a simple answer. We could cut taxes and raise transfer payments for the bottom three quintiles while raising taxes for the top two quintiles. But this is a pure zero-sum transfer. Every dollar given to the bottom 60 percent of the population must be taken away from the top 40 percent of the population. Politically this is exactly what we have been unable to do.

Jobs for the Middle Class

To determine the impact of government purchases on the mix of skills, the pattern of industrial demand, and the distribution of earnings, it is necessary to estimate total government employment and its characteristics. Total government employment includes direct government employees who work on government payrolls and indirect government employees who technically work for private companies but who produce government goods and services. From this perspective, the woodworker making school desks for a private furniture company is just as much a government employee as the schoolteacher on the public payroll. The furniture maker is simply an indirect government employee.

In 1976 governments (federal, state, and local) directly employed 18.4 percent of all those workers who worked. Of these 19.7 million individuals, 2.1 million were federal military employees,

3.4 million were federal civilian employees, and 14.2 million were state and local employees. Government employed 21 percent of all women, 16 percent of all men, 25 percent of all blacks, and 15 percent of all those of Spanish origin.[7]

If the occupational skill mix (see table 7–4) of government is compared with that of the private economy, there is one striking difference. Government, and especially state and local governments, provides employment for many more professional employees than the private economy does. Government employs 34.5 percent of all male professionals and 49.9 percent of all female professionals. When you consider all of the female schoolteachers, welfare workers, and nurses on public payrolls, this result is not as surprising as it first seems, but it means that highly educated women are heavily dependent on government expenditures for their job opportunities.

If an across-the-board cutback in government expenditures were to occur, workers would be laid off in proportion to their em-

TABLE 7–4

Percent of Occupational Employment Due to Direct Government Employment

	Federal Gov't.		State & Local Gov't.		Total Gov't.	
Occupation	Male	Female	Male	Female	Male	Female
Professional	7.1	2.9	27.4	47.0	34.5	49.9
Managerial	3.4	3.2	7.7	13.1	11.1	16.3
Sales	.2	.3	5.2	.3	5.4	.6
Clerical	16.7	5.5	10.9	16.8	27.6	22.3
Craftsmen	2.6	.6	5.4	4.7	8.0	5.3
Operatives, Excl. Transportation	.7	.4	1.1	1.1	1.8	1.5
Transportation Operatives	1.1	5.2	8.8	39.8	9.9	45.0
Nonfarm Laborers	2.2	1.7	1.8	4.4	14.0	6.1
Private Household	—	—	—	—	—	—
Service Workers Excl. Household	3.0	1.2	2.9	15.7	5.9	16.9
Farmers	—	—	—	—	—	—
Farm Laborers	1.1	—	.7	1.0	1.8	1.0
TOTAL	4.0	2.8	12.1	17.9	16.1	20.7

SOURCE: Extracted from the U.S. Bureau of the Census, *Current Population Reports, Consumer Income 1976,* Census tapes.

ployment in the government sector. As the government sector contracted, the private sector would expand, but expanding private demands would not employ the same skills released by government. If you examine table 7–4, any male occupational group that has more than 16 percent of its employment in government and any female group that has over 21 percent of its employment in government would find its job opportunities and wages diminishing.[8] Conversely, those who are underrepresented in government would find their job opportunities and wages rising as the private economy grew relative to the government economy.

Government employment directly alters the structure of employment opportunities, but it also directly affects the distribution of wages, since government does not necessarily pay the same wages as the private economy. One of the controversies about government employment is the extent to which it "over pays." The relative earnings of full-time, full-year government workers are presented in table 7–5. Basically, higher wages are paid to women and minorities. For white males the higher pay of the federal government is counterbalanced by the lower pay of the state and local government.

TABLE 7–5

Earnings of Government Workers Over and Above Those Received by Workers in the Private Economy (in Percent)

	Males	White	Black	Hispanic	Females	White	Black	Hispanic
Government	0	0	17	12	28	28	36	26
Federal	14	15	33	36	43	44	53	55
State & Local	–6	–6	9	2	25	26	31	19

SOURCE: Extracted from U.S. Bureau of the Census, *Current Population Reports, Consumer Income 1976,* Census tapes.

This leads to an argument as to whether government wages are higher because government hires more skilled employees, because government discriminates less against women and minorities, or because government simply pays more than it should. While it is difficult to settle the discrimination issue, it is relatively easy, due to the mix of skills employed, to determine the extent to which government wages are higher than those in the private economy.

This can be done by calculating the government wages that would be paid if every occupational skill in government were paid exactly what that same skill makes in the private economy. If government wages are higher than this calculation would indicate, the extra skills of the government labor force cannot explain its higher wages. When such a skill correction is made, the federal government's apparent overpayment to white males disappears. Government simply demands a higher-skilled labor force than the private economy. The same thing is true for white women in state and local government. Once a correction is made for their skills, their apparent overpayment also disappears. The apparent premium for white female federal workers, however, is reduced from 44 percent to 20 percent, but it is not eliminated.[9] Even after one corrects for skills, the federal government pays women substantially more than the private economy does. Minority females end up with an even larger premium (around 30 percent) in the federal government and a small premium (around 5 percent) in state and local governments. Minority males receive about a 20 percent premium in the federal government and are close to parity in state and local governments. Correcting for occupational skills increases the observant underpayment for white males in state and local governments since they end up earning only 82 percent as much as they would if they were paid the wages existing in the private economy. As a result, the question of overpayment really comes down to whether you believe that government discriminates less or whether it simply overpays. Whatever the truth, government raises the earnings of women and minorities above what they would be if only the private economy were to exist.

In addition to employing 18 percent of the labor force directly, government indirectly employs 11 percent of the private labor force through its purchases of goods and services from the private economy. But this percentage differs dramatically from industry to industry (see table 7–6).[10] Here again, any industry with more than 11 percent of its employment emanating from government would lose if government were cut back, and any industry with less than 11 percent of its employment attributable to government would gain as the private economy expanded. The large losers in the private economy would be construction, professional services, and durable goods manufacturing.

TABLE 7–6

Industrial Distribution of Employment in 1976

	Indirect Government (%)		Indirect Government as a Fraction of Private Employment (%)
	Federal	State & Local	
Agriculture	–	1.4	2.1
Mining	1.3	0.8	12.2
Construction	3.5	24.3	30.9
Durable Good Manufacturing	41.8	12.9	16.5
Nondurable Good Manufacturing	8.2	6.5	6.9
Transportation and Utilities	10.9	7.2	15.3
Trade	9.3	8.9	4.0
Finance	1.8	2.2	4.0
Business Services	5.1	4.7	13.4
Personal Services	3.9	5.5	9.0
Entertainment	1.7	0.4	6.2
Professional Services	12.6	25.3	17.2
Public Administration	–	–	–
Military	–	–	–
TOTAL	100	100	11.2

SOURCE: Calculated from U.S. Bureau of the Census, *Current Population Reports, Consumer Income 1976,* Census tapes and U.S. Department of Labor Input-Output Tables.

An examination of the distribution of earnings (direct plus indirect) generated by government reveals that it provides a greater proportion of jobs in the middle-income range, pays a higher-average wage, and generates a more equal distribution of earnings (see table 7–7).[11] If government were to disappear or to be proportionately cut back, the distribution of earnings would become more unequal. The earnings share of the bottom 60 percent of the work force would fall. The fourth quintile would break even and the richest quintile would raise their share of total earnings.

For the middle class, cuts in government expenditures are counterproductive. Their taxes would go down, but their income would go down even faster. In the end they would have a lower real standard of living. Whatever they think, they are one of the prime beneficiaries of government in both direct and indirect employment.

TABLE 7–7
Private Versus Public Earnings

Quintile (Poorest to Richest)	Government (%)	Private * (%)
1	2.9	2.3
2	9.8	7.4
3	20.3	15.7
4	25.3	25.4
5	41.7	49.1
Mean earnings	$9,553	$8,431
Number of Workers (million)	29.4	77.6

* Refers to that part of private employment generated by private demands.
SOURCE: Calculated from U.S. Bureau of the Census, *Current Population Reports, Consumers Income 1976,* Census tapes and U.S. Department of Labor Input-Output Tables.

Who Has the Tax Loopholes

Taxation requires explicit equity decisions. Income must be taken from someone to finance government expenditures, but whose income should go down? There is no economic answer to this question. An answer can only flow from your vision of what constitutes a just distribution of after-tax income.

While taxation requires equity decisions, nowhere is our inability to make equity decisions more in evidence. Internal contradictions abound. Progressive and regressive taxes coexist. Often highly progressive nominal rates hide regressive effective rates once the effects of special provisions are taken into account. The loopholes in the federal income tax are well-known, but they are matched by even larger, but less well-known, loopholes in every other tax we have. Periodic popular demands for progressive reforms—usually just after the government has announced the number of millionaires (twenty-four) that pay absolutely no taxes—coexist with the absence of actual reforms.

The easiest explanation for these contradictions would be to argue that they are created by the political desire to do something that is economically unfeasible. We want to have a more equal distribution of economic resources, but the need to maintain work incentives prevents us from achieving this goal. Unfortunately this is an argument that has long been breached in the economics literature even if it has not been heeded in public rhetoric. Work incentives are important and it is possible to impose such high taxes that they interfere with work effort; but all of our empirical studies show that our current taxes are far below the levels that create disincentives to work.[12] The highest marginal tax rate on earnings is now just 50 percent. Repeated studies have shown that highly progressive tax systems (much more progressive than the tax system now in place) do not seem to reduce work effort. Income effects (the need to work more to regain one's living standards) dominate substitution effects (the desire for more leisure because of lower take-home wage rates), and individuals work for a variety of other rewards—power, prestige, promotions, satisfaction.

The tensions and contradictions in our tax system are generated not by "economic necessity" but in ourselves. Despite a recurring interest in progressive tax reforms, general tax reform programs have consistently failed to win political and perhaps popular approval. There are two reasons for this: one has to do with our definition of "the rich" and the other has to do with our conception of "individual merit."

Discussions of "the rich" tend to talk about income rather than wealth. Since many of the returns from wealth are not counted as income (unrealized capital gains being the most notable), inequalities are reduced and the rich are perceived as less rich than they actually are. While the top quintile of all households has almost 80 percent of total wealth, it has only 44 percent of total income according to census definitions. Correcting census definitions to reflect all capital income raises the income share of the top 1 percent of all households from 5 percent to 11 percent of total income.[13]

But focusing on income leads to a more fundamental problem than the simple generation of misleading statistics. While the top quintile of all households has 44 percent of total income, most members of this group do not think of themselves as rich, and what is more important, they are not thought of as rich by the rest of society. In 1977 an income of $24,000 places a household in the top quintile.[14] Many households are in the top quintile because of the effort of two workers. A husband and wife each earning $12,000 belong to the top quintile, but are they rich? Neither they nor their neighbors think so. The top 5 percent of all families have 17 percent of total income, but only $38,000 is necessary to place a family in this exclusive class. A husband and wife each earning $19,000 qualify. Many would be willing to argue that even this family is not rich. They are simply a hard-working, middle-class family.

But once you get up to income levels where there would be general agreement that the family is rich, there is so little taxable income, as it is officially defined, that it is impossible to promise substantial income tax reductions for the rest of the population by raising the tax rates of the rich. To tax the rich it is necessary to change the official definitions of income, but this involves closing loopholes that provide small dollar benefits to the middle class, although most of the benefits may go to the rich. The middle class sees the loopholes that it is going to lose and fears that it will not get an equally large general tax cut in return. The middle class's ambivalence to tax reform creates a demand for tax reform in the abstract but not enough political force for any concrete reform proposals to become law.

Widely held conceptions of "individual merit" also impose limitations on the income tax-transfer system. Most individuals think of their earnings as something that accrues to them as a matter of personal merit and productivity. Being a meritorious award, they can think of little reason why some of it ought to be taken from them and given to others.

The net result is a tax system that is a hodge-podge of progressive, regressive, and proportional taxes, and with many taxes there is a great uncertainty as to who pays them. This uncertainty is greatest in the corporate income tax since it can be viewed as either a tax on the rich (the top ½ of one percent of the population own 43 percent of all corporate stock) or a sales tax on the middle class and poor who buy corporate goods. Depending upon exactly how the corporate income tax is treated, estimates of the incidence of the current tax system can vary from being quite progressive to quite regressive. Whatever its absolute incidence, however, the tax system is becoming more regressive as we raise the regressive taxes (sales and payroll) and lower the progressive taxes (income).

The truth about our tax system, however, is not that it is progressive or regressive, but that it is unfair. Many high-income individuals pay little or no tax; many others pay high taxes. Some low-income individuals pay high taxes; most do not. This is true regardless of whether you regard the tax system as progressive or regressive. Individuals with exactly the same income end up paying very different taxes in the United States.

Table 7–8 presents one estimate of the distribution of tax burdens. To understand who pays taxes it is necessary to look not just at the average tax collections for each income class but the variation within that class. According to these estimates the average tax rate rises from 16.8 percent for the poorest 10 percent of the population to 26.2 percent for the richest 10 percent of the population.[15] But around these averages there is an enormous spread. The second column presents what are called the standard deviations around these mean values. Suppose that you were looking at the fifth decile with an average tax rate of 22.8 percent and a standard deviation of 6.5. This says that 68 percent of the taxpayers in this income class

TABLE 7–8
Variance in Tax Rates

Deciles of the Population	Mean Tax Rate (%)	Standard Deviation
1	16.8	30.1
2	18.6	14.6
3	21.6	19.6
4	22.6	8.8
5	22.8	6.5
6	22.7	5.5
7	22.7	6.6
8	23.1	5.9
9	23.2	5.4
10	26.2	10.2

SOURCE: Joseph A. Pechman and Benjamin A. Okner, *Who Bears the Tax Burden?* (Washington, D.C.: The Brookings Institution, 1974), p. 67.

would be included in an interval from 16.3 percent (22.8 – 6.5) to 29.3 percent (22.8 + 6.5). To include 95 percent of the taxpayers you would have to have an interval of two standard deviations (22.8 +/– 13.0), and to include 99 percent of all taxpayers, you would have to have an interval of three standard deviations (22.8 +/– 19.5). With such wide ranges needed to include most of the taxpayers in any income class, average tax rates mean very little. They do not tell us what the average person pays. The truth about our tax system is not progressivity or regressivity, but dispersion—the unequal treatment of equals.

If the current demands for tax cuts on capital income to accelerating economic growth were to be met, this situation would become much worse. Whatever the current degree of progressivity in the tax system, it would be lessened, and whatever the current degree of horizontal inequity, it would be magnified. Given that there is a potent political demand to cut taxes on capital income, it is worth spending some time looking at the generation and taxation of capital income.

This is an area where there is one legitimate issue and a host of illegitimate ones. The legitimate issue is that of inflation. With inflation, investors often must pay taxes on capital gains that are not real capital gains. This should be corrected with an indexed tax system that only taxes real income gains. But the same system should also exist for wage earners. They also get taxed on money rather than real income gains. Indexation should exist for both wage earners and capital investors.

As we have already seen, capital investment should be encouraged by abolishing the corporate income tax. Once this has been done and the issue of inflation indexing has been surmounted, there is no case for further cuts in personal taxes on capital income. When anyone starts talking about restructuring taxes to encourage saving and investment, it is well to remember that our tax system already taxes capital very lightly, if at all. The great current loophole in American taxation is the fact that great wealth can be generated, controlled, spent, and passed on to one's children without ever being subject to the levels of taxation faced by modest wage earners. Discussions of loopholes often imply that there are loopholes for the rich and not for the poor; this is inaccurate. There are many tax loopholes for capitalists, rich or poor, and few tax loopholes for wage earners, rich or poor. Given our professed attachment to the ideal of hard work, this is perhaps surprising, but nevertheless true. One of the major causes of this problem is a system of income taxation that doesn't succeed in taxing the main path to wealth. In the mythology of getting rich, a poor man starts with some initial earnings power. Out of these earnings he patiently saves, pays taxes, invests, pays taxes, saves, and reinvests at market rates of interest until he becomes wealthy. The only trouble with this model is that it does not describe how individuals actually become rich.

If one examines the very rich, about 50 percent of the great fortunes are gotten through inheritance. Despite what we often hear about so-called confiscatory inheritance taxes, U.S. gift and inheritance taxes amount to a tax of only 0.2 percent on net worth.[16] For all practical purposes, the current estate and gift tax system has no impact on the distribution of wealth. If you are very rich and want to hand it on to your children, nothing stops you from doing so.[17]

The more interesting half, however, is the half that did not inherit their fortunes. These individuals accumulate so much and so rapidly that it could not possibly have come through a process of patient savings, taxation, and investment at market rates of interest. A dramatic illustration of this phenomenon is a *Fortune* article that listed thirty-nine individuals who had made from $50 to $700 million in the previous five years without inheriting wealth or having previously been on the *Fortune* lists of the wealthiest Americans.[18] The prevalence of instant wealth is also visible if one looks at the names of those who inherited great wealth. The Rockefellers, Mellons, Fords, Whitneys, and Posts may have inherited their wealth, but these fortunes were made very quickly at some point in the past.

To understand how this is done, it is necessary to think in terms of two different capital markets. The first capital market is the market for physical investments. In this market, firms and individuals make real investments in plants and equipment. The second capital market is the financial market where individuals buy financial instruments (ownership rights) without directly managing real plant and equipment. Stocks, bonds, and real estate trusts are examples of the latter; factories, stamping presses, and lathes are examples of the former.

Instant wealth arises in the process of capitalization. Consider a real investment in plant and equipment of $10 million that earns $3 million per year. Suppose that the market rate of interest or the discount rate is 10 percent. With a 10 percent discount rate a $3 million annual income flow is worth $30 million ($3 million/.10) in the financial market. If the discount rate were 5 percent, the same investment would be worth $60 million. This is true regardless of how much it cost to make the initial investment. But in our example, the initial investor has now increased his wealth instantly from the initial $10 million to $30 million when the investment was sold, The purchaser who buys the stock for $30 million, however, has an investment that only earns the market rate of interest (10 percent). If the real investment opportunity were something that could be duplicated by the initial investor, his $10 million investment might be worth even more because of the future profits that similar investments could bring.

It is this process of capitalizing above-average returns that generates rapid fortunes. Patient savings and reinvestment have little or nothing to do with such fortunes. To become very rich one must generate or select a situation where an above-average rate of return is about to be capitalized.

If real capital markets reached equilibrium quickly, large fortunes could not be made in this process of capitalization. Once a new physical investment opportunity was discovered, real investment funds would quickly flow into the area and bring the real rates of return down to the market's average rate of return. Above-average profits would not be expected to last very long, and there would be no possibility of obtaining a monopoly on future above-average physical investment opportunities. Other people would move into the area and future physical investments would only earn the market rate of return. In this case, physical investments are only worth what they cost to build and cannot cause sudden additions to wealth.

Data on real capital markets indicate little if any tendency for real capital markets to approach equilibrium. Substantial, persistent differences in real rates of return exist. The reasons for this fundamental disequilibrium are many and varied, but most of them spring from a basic characteristic of real investment markets. Investment resources simply do not flow quickly across firms and industries thereby equalizing real rates of return.

Most of our savings are not personal savings but consist of the internal savings of businesses in the form of retained earnings and depreciation allowances. But firms almost always reinvest their own internal funds and seldom invest in other firms even if their rate of return on investments is far below the national average.

To explain why internal funds are frozen into the firm generating them it is only necessary to think about the basic characteristics of U.S. capitalism. It is managerial capitalism. Large firms are controlled by individual managers who usually do not own any substantial fraction of the firm that they manage. While a stockholder might like to see his funds invested in the highest rate of return industries, regardless of who manages these industries, the existing manager clearly has other incentives. He wants to use internally generated savings for investments under his management, since this is the pattern of investments that brings him increasing returns in the form of income, power, and prestige. As a result, those who direct real investments are not simply profit-maximizing investors. They are interested in maximizing profits, but only profits from operations that they themselves manage.

If we ask why managers with large internal savings do not start subsidiaries in high-profit industries rather than reinvesting in their own low-profit industries, we come face to face with the entire structure of restricted competition in the U.S. economy. Barriers to entry are often high, and managers often do not have the specialized knowledge necessary to make profits in another industry. The existence of high profits in the cosmetics industry, for example, does not mean that iron and steel executives could earn high profits there. True, the firm might be able to earn high profits, but it would have to fire its existing managers and hire new managers. The existing managers are not about to fire themselves, and they are wise enough to know, or have learned via the unsuccessful conglomerate movement, that they could not run a successful cosmetics firm. As a result, they stay in the steel industry and reinvest their internal funds in steel regardless of the relative rates of return.

The existence of internal savings also tends to distort the flows of those few investment funds that do flow through the real capital market. In the real world, lenders face risks and uncertainties about actual returns. If they lend to firms with large flows of internal savings, they can have great confidence that borrowers are going to repay their loans regardless of the success or failure of the actual project for which the funds were lent. Because of the low risk of default, funds are attracted to those firms with large internal savings, regardless of whether they are earning above-average rates of return on their capital investments. The net result is a flow of external savings that does not serve to equalize real rates of return across the economy either.

While the process of capitalizing disequilibrium rates of return explains instantaneous wealth, there is still the problem of how these fortunes are allocated to individuals. This brings us to what is called the *random walk*. Since no one can predict where these opportunities for capitalizing real disequilibriums out of existence will appear, the winners are, as in any lottery, lucky rather than smart or meritocratic.

For example, in the early 1950s you might have invested in a class of firms that included Xerox. In 1950 all of these firms would have looked alike and all would have had an equal expected rate of return. Looking back, some would have gone broke and disappeared; most would have earned the market rate of return; some would have earned more than the market rate of return; and a few, perhaps one, would have been an investment such as Xerox. Those who owned shares in it became wealthy. They won the lottery.

The random walk is a process that will generate a highly skewed distribution of wealth. You cannot lose more than you have, but you can make many times what you have. Because most holders of wealth eventually diversify their portfolios, great fortunes remain even if the underlying disequilibrium in the real capital market eventually disappears. It should be emphasized that there is no equalizing principle in the random walk. Those who have had good luck are no more apt than the random individual to be subject to bad luck.

What is the evidence for the random walk hypothesis? [19] First, an examination of large financial firms (such as mutual funds) indicates that none of them is able to outperform the market averages. Professional financial managers able to make large investments in obtaining market information are not able to outperform the market average or a random drawing of stocks. Second, no one has been able to design a set of decision rules (when to buy and sell) that yields a greater than average rate of return. Third, tests indicate that stock prices quickly adjust to changes in information (announcements of stock splits, dividend increases, and so forth). Fourth, there is no serial correlation among stock prices over time. The price at any moment in time or its history cannot be used to predict future prices. When put together, all of these findings form an impressive body of evidence as to the existence of the random walk.

While many of the great fortunes represent a combination of entrepreneurial and financial investments, the same random walk process probably holds. Ability is necessary, but within a group of individuals with equal entrepreneurial talents a nonnormal random lottery occurs. There is an expected rate of return for the group as a whole, but there exists a wide dispersion in individual results around this average. Entrepreneurial talent is a necessary condition for entering the lottery, but it is not a sufficient condition for making instantaneous wealth. The entrepreneurial random walk is, however, much less subject to proof than the purely financial variant. The unsuccessful entrepreneur does not remain visible for study in the same manner as the unsuccessful stockholder.

The net result is a process that generates a highly skewed distribution of wealth. Great wealth is created in relatively short periods of time. Personal savings behavior and one's ability to postpone future gratification have little or nothing to do with the process. Once created, large fortunes maintain themselves either because the underlying disequilibrium in real returns remains or because investments are diversified and earn the market rate of return.

If you read the *Fortune* biographies that accompany their lists of the most wealthy, the winners will be described as brighter than bright, smarter than smart, quicker than quick.[20] But look beyond the description to see if they were simply lucky or possess some unique abilities. Remember that the unsuccessful entrepreneur of equal ability will not be featured in *Fortune*. To what extent were they like many other people but in the right place at the right time? The real test of unique abilities is to ask how many have repeated their performance. How many have made a great fortune on one activity or investment and then managed to go on to earn another great fortune on another activity or investment? If the *Fortune* list is examined, it is impossible to identify anyone whose personal fortune was subject to two or more upward leaps. The typical pattern is for a man to make a great fortune and then settle down and earn the market rate of return on his existing portfolio.

What has been generated in this process is realized and unrealized capital gains. Realized capital gains are taxed at less than half of normal rates, and unrealized capital gains are not taxed at all. Those thirty-nine, five-year multimillionaires in the *Fortune* article undoubtedly paid little or no taxes. Nor should one imagine that it is impossible to consume unrealized capital gains without paying taxes. Simply go to your friendly banker (if one is a multimillionaire, there are many friendly bankers), take out a loan using your appreciated stock as collateral, and buy whatever you like. The interest payments on the loan are even tax deductible. You can consume whatever you like and pay no taxes. At death, the principal can be repaid out of that same appreciated stock.

Since World War II, our three-part system has been extremely viable—income transfers for the poor, direct or indirect government jobs for the middle class, and little or no taxation for wealthy capitalists. But cracks are appearing in the system like those cracks in the wings of the DC-10. Because of female work patterns, we are probably entering a period of rising income inequality where the second and third quintiles of the population are going to lose ground in their earnings. This, coupled with the view that inflation could be cured if only government expenditures were cut, will undoubtedly lead to an environment where transfer payments do not rise rapidly enough to hold the poor and the elderly even with the rest of the population. If government expenditures are actually cut or fall with falling school enrollments, income pressures on the middle class will mount. At the same time, demands to cut taxes on capital income in the name of accelerating economic growth will raise the tax burden and lower the income share of those who are not capitalists.

Group Demands

In our society the whole issue of group justice is often seen as illegitimate. Individual blacks may have been unfairly treated, but blacks have not been treated unfairly as a group. Consequently, remedies must come at the individual level (a case-by-case fight against discrimination or remedial education programs for individuals) and not at a group level. Affirmative action or quotas programs that create group preferences are fought on the ground that they are unfair even if everyone agrees that many or all members of the group to be helped have suffered from unfair treatment in the past.

Our economic theory is based upon the same tradition. Western economics is at its heart an economics of the individual. Individuals organize voluntary economic associations (the firm), but individuals earn and allocate income. Group welfare is, if anything, only the algebraic summation of the individual welfare of the members of the group. There are no involuntary groups. Individuals join groups only when groups raise individual welfare. No one assigns someone to a group to which he or she does not wish to belong. Race and sex are not economic variables from this perspective.

At the same time, our age is an age of group consciousness. Economic minorities argue that group parity is a fundamental component of economic justice and that an optimum distribution of income consists of more than an optimum distribution of income across individuals. In doing so, they are not advocating something new but extending to themselves old doctrines that are invoked to help farmers and many other industries. While there is plenty of precedent for helping groups in our economy, a faster rate of growth may call for ending help to different industrial groups. Should the same principles be used to resist instituting aid for other social groups? Is the correct economic strategy to resist group welfare measures and group redistribution programs wherever possible? Or do groups have a role to play in economic justice?

We are a society that professes belief in "equal opportunity" for individuals, but how could you tell whether equal opportunity does or does not exist? In a deterministic world we could tell whether equal opportunity existed by seeing whether each individual reached a level of economic performance consistent with his or her inputs (talents, efforts, human capital). Individuals could be identified as receiving less than equal treatment.

But the real world is highly random and not deterministic. Since everyone is subject to a variety of good and bad random shocks, no one can tell whether any individual has been unfairly treated by looking at his or her income. Individuals may have participated in the same economic lottery, but in the end, someone lost and someone won. My low income and your high income do not prove that I was unfairly treated relative to you. You were lucky and I was unlucky. But I was not unfairly treated, and I did not suffer from discrimination or some systematic denial of equal opportunities.

Since those variables that we normally think of as the deterministic variables—education, skills, age, and so forth—only explain 20 to 30 percent of the variance in individual earnings, our economy is one where the stochastic shocks (or unknown factors) are very large relative to the deterministic (or known) part of the system.[21] And the larger the stochastic portion of the system relative to the deterministic portion of the system, the less possible it is to identify individuals who have been unfairly treated. In the economic area, no one can say that any individual has been subject to systematic discrimination as opposed to random bad luck. This is a judgment that can only be made at the level of the group.

This can be seen in the standard economic tests for the existence of discrimination. Earnings data are collected for different groups of individuals, and a statistical equation is estimated to show the relationship between earnings and the normal human capital factors (work effort, skills, education) for each of the groups. These equations are then examined to see if they are significantly different. If they are, the different groups do not participate in the same economic lottery.

Using economic analysis it is impossible to determine whether any individual has suffered from the denial of equal opportunity. Within any group—no matter how privileged—there will be individuals who have been denied equal opportunities and suffered from discrimination, but they have not been subject to a systematic denial of opportunities. Society may be concerned, but it is completely incapable of doing anything about random discrimination. It is simply one type of random good or bad luck that affects us all. A Polish American may feel aggrieved and may have been denied equal opportunities, but Polish Americans do not suffer from systematic denials of equal opportunity since their earnings functions do not meet the necessary tests. Conversely, within any group—no matter how underprivileged—there will be individuals who have not suffered from a systematic denial of opportunities.

All society can do is to test whether the economic lottery played by whites is or is not statistically equivalent to the economic lottery played by blacks. It cannot tell whether any individual, black or white, has been equally treated. Discrimination affects individuals, but it can only be identified at the level of the group. As a result, it is not possible for society to determine whether it is or is not an equal opportunity society without collecting and analyzing economic data on groups.

But the measurement problem also creates a remedy problem. If it is impossible to identify individual discrimination, upon whom should the remedies for systematic discrimination be focused? The inability to identify anything except group discrimination creates an inability to focus remedies on anything other than the group. We can attempt to create an economy where everyone participates in the same economic lottery, but we cannot create an economy where each individual is treated equally. If you believe current earnings functions, 70 to 80 percent of the variance in individual earnings are caused by factors that are out from under the control of even perfect government economic policies. The economy will treat different individuals unequally no matter what we do. Only groups can be treated equally.

Groups, rather than individuals, are also going to enter into our decisions because we need groups to make efficient decisions. At the same time, what is "efficient" for the economy is always "unfair" to some individuals. The problem is now to balance the gains from efficiency against the losses from unfairness.

Suppose that you were the dean of a medical school charged with the task of maximizing the number of M.D.'s produced for some given budget. In the process of carrying out this mandate, you noticed that 99 percent of all male admissions completed medical school, and that 99 percent of all male graduates go on to become lifetime doctors, but that the corresponding percentages for women were each 98 percent. As a consequence, each male admission represents .98 lifetime doctors and each female admission represents .96 lifetime doctors. Seeking to be efficient and obey your mandate to maximize the number of practicing doctors, you establish a "male only" admissions policy.

In this case, the dean of the medical school is practicing statistical discrimination. He is treating each group fairly, based on the objective characteristics of the group, but he is unfairly treating 96 percent of all women because they would, in fact, have gone on to become practicing doctors. His problem is that he has no technique for identifying which 4 percent of all women will fail to become practicing doctors, and therefore he expands a very small difference in objective characteristics (a one percentage point difference in each of the two probabilities) into a zero-one decision rule that excludes all women. Is the dean acting fairly or unfairly, efficiently or inefficiently?

To be efficient is to be unfair to individuals. Where is the balance to be drawn? Wherever the balance is drawn, groups become important since it is efficient for employers to open or close opportunities to individuals based on the groups to which employers assign them. But since employers will of necessity use groups, government must become involved in the question as to what constitutes a legitimate group or an illegitimate group. The option of prohibiting all decisions based on group characteristics simply isn't possible since the efficiency price would be too high.

A controversy of just this type recently arose in Massachusetts over automobile insurance rates. In the past, these rates have been set based on the age, sex, and geographic location of the driver and the associated accident data. The insurance commissioner of the state shifted to a system that rates drivers based upon the number of years they have had a license, their accident record, and their arrest record. Different individuals will pay very different insurance premiums under the two systems. Which is the right set of groups?

Ideally, group data could only be used for making economic decisions where all members of the group had the same characteristics. Fair treatment for the group would be a fair treatment of each individual member of the group. Unfortunately, this situation almost never exists. Homogeneous groups do not exist. A trade-off must be made between efficiency and justice. Since employers are only interested in efficiency, they will make the trade-off in favor of efficiency and in favor of unfair individual treatment unless they are restrained from using certain group classifications. As a result, the state is forced to establish categories of illegitimate groups. Our social desires for individual justice, at least to some extent, take precedence over our social desires for efficiency.

Since we have both a desire for efficiency and a desire for individual justice, we have a dilemma. Individuals have to be judged based on group data, yet all systems of grouping will result in the unfair treatment of some individuals. Thus we must establish some standard as to how large differences in mean characteristics have to be before a particular set of groups is legitimate. Most of us would be unwilling to let the dean of our medical school exclude women on the basis of a 1 percent difference in objective probabilities, but what would our judgments be if the objective differences were fifty percentage points or ninety percentage points? At what point would we be willing to exclude women? Yet if we did this at any point, we would be unfairly treating some individual female. But if we did not exclude them, we would be wasting a larger and larger fraction of our resources.

What this illustrates, however, is that every society has to have a theory of legitimate and illegitimate groups and a theory of when individuals can be judged on group data and when they cannot be judged on group data. A concern for groups is unavoidable.

On first thought, mobility (or the lack of mobility) would seem to be an easy way to determine what groupings are legitimate. If it is easy for an individual to leave any group, then individuals in that group cannot claim to be unfairly treated. The value of the group must exceed the costs of the group or they would not belong. They may receive less measurable income by being a member of the group, but their psychic income from being a member of the group must at least counterbalance the lower measurable income. It is precisely this argument that lies at the heart of the recommendation that government should not have special programs to raise the money incomes of farmers. Farmers may have lower incomes than urban dwellers, but they could always cease to be farmers and become urban dwellers. Therefore, farmers cannot be unfairly treated regardless of the relative income of farmers and regardless of the sources of this relative difference.

While this argument may sound reasonable to those of us who are not farmers, it is equally applicable to regions or religions. Technically it is just as easy, if not easier and less costly, to move from one region to another or from one religion to another. Yet most of us would not be willing to argue that one must change his or her religion to achieve economic parity. Why? What is the difference between changing one's occupation and one's religion? Individuals can certainly be just as psychologically committed to a particular occupation as they are to a particular religion.

If one looks at our social programs, society certainly cannot claim to focus consistently on individuals as opposed to groups. Affirmative action for economic minorities may be on the defensive, but we are in an age when industrial and regional programs are expanding rapidly. The same people who oppose special programs for blacks support special programs for textiles. Imagine the furor that would arise if we started a program for blacks similar to that now in place for farmers. It would be denounced as "un-American" from every rooftop. Given that our society clearly is not willing to be consistent and use an individual focus when it comes to politically popular groups, it is easy to see why the insistence on an individual focus for minorities can be viewed as simply a more sophisticated version of discrimination. Those who got ahead in the economic race stay ahead for a very long time even after discrimination has ceased to be actively practiced.

While we undoubtedly are not willing to use mobility as the sole test of whether a group is illegitimate—almost no one would be willing to force individuals to change their religion to secure equal economic treatment—it is still the basic ingredient. We need to be most concerned about discrimination against groups where individuals cannot easily leave the group in question. In the case of industrial groups, this means that society should focus on improving the ease of exit for individuals rather than aiding the group as it is now constituted. When ease of exit is high, we can concentrate on efficiency, knowing that it will not lead to much unfairness. When the ease of exit is low, the reverse is true.

In the end we have a problem. Various groups are demanding parity in their income position and there is little reason to believe that these demands will disappear in the future. Few are willing to stand up and publicly defend the idea that blacks, Hispanics, and women should permanently earn less than white males. There are many who object to every conceivable remedy, but this only exacerbates the tension without either solving the problem or causing it to go away.

Nor will the normal actions of the economy cause the problem to fade away with time. A simple look at what has been happening can force anyone to abandon the comfortable "do-nothing" hypothesis. The essence of any minority group's position can be captured with the answer to three questions: (1) Relative to the majority group, what is the probability of the minority's finding employment? (2) For those who are employed, what are the earnings opportunities relative to the majority? (3) Are minority group members making a breakthrough into the high-income jobs of the economy? In each case, it is necessary to look not just at current data, but at the group's economic history. Where has it been, where is it going, how fast is it going, and how fast is it progressing?

In terms of ethnic origin, there are three economic minorities in the United States—blacks, Hispanics, and American Indians. Of the almost 100 million other Americans who list themselves in the census as having an ethnic origin, all have incomes above those of Americans who list no ethnic origin. The highest family incomes are recorded by Russian–Americans, followed by Polish–Americans and Italian–Americans.[22] "Ethnic" Americans sometimes talk as if they were economically deprived, but they are actually perched at the top of the economic ladder. Females constitute the other major economic minority. Many of them may live in families with high incomes, but when it comes to earnings opportunities, they do not participate in the same economic ball game as men.

If you examine the employment position of blacks, there has been no improvement and perhaps a slight deterioration. Black unemployment has been exactly twice that of whites in each decade since World War II. And the 1970s are no exception to that rule. Whatever their successes and failures, equal opportunity programs have not succeeded in opening the economy to greater employment for blacks. Given this thirty-year history, there is nothing that would lead anyone to predict improvements in the near future. To change the pattern there would need to be a major restructuring of existing labor markets.

Viewed in terms of participation rates, there has been a slight deterioration in black employment. In 1954, 59 percent of all whites and 67 percent of all blacks participated in the labor force. By 1978 white participation rates had risen to 64 percent and black participation rates had fallen to 63 percent.[23] This change came about through rapidly rising white female participation rates and falling participation rates for old and young blacks. In the sixteen to twenty-one age category, black participation rates are now fifteen percentage points below that for whites.

At the same time, there has been some improvement in the relative earnings for those who work full-time, full-year. In 1955 both black males and females earned 56 percent of their white counterparts. By 1977 this had risen to 69 percent for males and 93 percent for females.[24] While black females made good progress in catching up with white females, this has to be viewed in a context where white females are slipping slightly relative to white males. If black males were to continue their relative progress at the pace of the last twenty years—five percentage points every ten years—it would take black males another sixty years to catch up with white males.

While the greatest income gains have been made among young blacks and one can find particular subcategories that have reached parity (intact college-educated, two-earner families living in the Northeast), there still is a large earnings gap among the young. Black males twenty-five to thirty-four years of age earned 71 percent of what their white counterparts earned in 1977. Among full-time, full-year workers, the same percentage stood at 77 percent.[25] Young black males are ahead of older black males, but they have not reached parity. As with black females in general, young black females do better than males. Females twenty-five to thirty-four years of age earned 101 percent of whites, and full-time, full-year black females earned 93 percent of what whites earned.

Using the top 5 percent of all jobs (based on earnings) as the definition of a "good job", blacks hold 2 percent of these jobs while whites hold 98 percent.[26] Since blacks constitute 12 percent of the labor force they are obviously underrepresented in this category. Relative to their population, whites are almost seven times as likely to hold a job at the top of the economy than blacks. At the same time, this represents an improvement in the position of blacks relative to 1960. Probabilities of holding a top job have almost doubled.

Separate data on Hispanics only started at the end of the 1960s and is not as extensive as that available for blacks, but during the 1970s Hispanics seemed to have fared slightly better than blacks in the labor market. Where their family income was once lower than that of blacks, it is now higher. This is probably due to the fact that Hispanics are much more heavily concentrated in the sunbelt, with its rapidly expanding job opportunities.

Instead of having unemployment twice that of whites, unemployment is only 45 percent higher.[27] Labor force participation rates are rising even more rapidly than those for whites. In terms of relative earnings, full-time, full-year males earn 71 percent of what whites earn, and females have reached 86 percent of parity.[28] While there are substantial differences in family income among different Hispanic groups, earnings are very similar among the major groups. In 1976 Cuban–Americans, Mexican–Americans, and Puerto Ricans were all within $200 of each other in terms of personal income, for those with income.

In terms of the best jobs, Hispanics hold 1 percent of these jobs but constitute 4 percent of the labor force. Relative to their population, whites are three times as likely to be in the top 5 percent of the job distribution as Hispanics.[29] In terms of breaking into the good jobs of the economy, Hispanics are far ahead of blacks.

Direct Redistributional Issues

American Indians are the smallest and poorest of America's ethnic groups. They are poorly described and tracked by all U.S. statistical agencies. Despite the existence of the Bureau of Indian Affairs, only the roughest estimate for their economic status is available. In terms of family income, reservation Indians probably have an income about one-third that of whites. Where nonreservation Indians stand no one knows.

Female workers hold the dubious distinction of having made the least progress in the labor market. In 1939 full-time, full-year women earned 61 percent of what men earned.[30] In 1977 they earned 57 percent as much. Since black women have gained relative to black men, white women have fallen even more in relation to white men over this forty-year period. Adult female unemployment rose from 9 percent higher than men in 1960 to 43 percent higher in 1978. From 1939 to 1977 the percentage of the top jobs held by females has fallen from 5.5 percent to 4 percent although women rose from 25 percent to 41 percent of the labor force. Relative to their population, a man was seventeen times as likely as a woman to hold a job at the top of the economy in 1977.

With the exception of breaking into the top jobs in the economy, much of this decline can be attributed to rapidly rising female participation rates. With more women in the labor force, there is simply more competition leading to lower wages and more unemployment. At the same time, the results indicate that the structure of the economy has not changed, and women have not broken through into a world of equal opportunity. In such a world they would compete with men and not just with each other.

At the bottom of the labor force stand the young—our modern lumpen proletariat. In 1978, 49 percent of all unemployment was concentrated among sixteen- to twenty-four-year-olds.[31] Unemployment rates were three times that of the rest of the population. Among male full-time, full-year workers, relative earnings stood at 40 percent for fourteen- to nineteen-year-olds and 65 percent for twenty to twenty-four-year-olds.[32] Among females the same percentages were 64 and 104. In terms of holding the top jobs, sixteen- to twenty-four-year-olds held 0.5 percent although they constituted 24 percent of the labor force.

While low earnings can be dismissed on the grounds that the

group is acquiring skills and will in the future earn higher incomes, the unemployment is not so easy to dismiss. Unemployed young people or young people who have dropped out of both school and the work force represent individuals who are not acquiring skills and good work habits. What this portends for the distribution of earnings in the future is hard to say since we have never before had a period where so much of the unemployment of our society is concentrated among the young and especially among young minorities. Certainly it is hard to think that it will do anything except make the distribution of earnings more unequal in the future.

While it is convenient to the position that if we were just to eliminate discrimination and create an equal opportunity world, minority group problems would take care of themselves, this position is untenable in both practice and theory. Imagine a race with two groups of runners of equal ability. Individuals differ in their running ability, but the average speed of the two groups is identical. Imagine that a handicapper gives each individual in one of the groups a heavy weight to carry. Some of those with weights would still run faster than some of those without weights, but on average, the handicapped group would fall farther and farther behind the group without the handicap.

Now suppose that someone waves a magic wand and all of the weights vanish. Equal opportunity has been created. If the two groups are equal in their running ability, the gap between those who never carried weights and those who used to carry weights will cease to expand, but those who suffered the earlier discrimination will never catch up. If the economic baton can be handed on from generation to generation, the current effects of past discrimination can linger forever.

If a fair race is one where everyone has an equal chance to win, the race is not fair even though it is now run with fair rules. To have a fair race, it is necessary to (1) stop the race and start over, (2) force those who did not have to carry weights to carry them until the race has equalized, or (3) provide extra aid to those who were handicapped in the past until they catch up.

While these are the only three choices, none of them is a consensus choice in a democracy. Stopping the race and starting over would involve a wholesale redistribution of physical and human wealth. This only happens in real revolutions, if ever. This leaves us with the choice of handicapping those who benefitted from the previous handicaps or giving special privileges to those who were previously handicapped. Discrimination against someone unfortunately always means discrimination in favor of someone else. The person gaining from discrimination may not be the discriminator, but she or he will have to pay part of the price of eliminating discrimination. This is true regardless of which technique is chosen to eliminate the current effects of past discrimination.

An individualistic ethic is acceptable if society has never violated this individualistic ethic in the past, but it is unacceptable if society has not, in fact, lived up to its individualistic ethic in the past. To shift from a system of group discrimination to a system of individual performance is to perpetuate the effects of past discrimination into the present and the future. The need to practice discrimination (positive or negative) to eliminate the effects of past discrimination is one of the unfortunate costs of past discrimination. To end discrimination is not to create "equal opportunity."

The problem of group demands cannot be left to the economy to solve. Major elements of the problem are not being solved at all and where progress is being made it is so slow that economic minorities would have to be patient for many more years. Yet any government program to aid economic minorities must hurt economic majorities. This is the most direct of all of our zero-sum conflicts. If women and minorities have more of the best jobs, white males must have fewer. Here the gains and losses are precisely one for one.

The Paradigm Zero-Sum Game

When society has to confront the issue of differences in the relative income of different groups–rich versus poor, black versus white, male versus female, farmers versus urban dwellers–it is addressing the paradigm zero-sum game. Every increase in the relative income of one group is a decrease in the relative income of another group. The gains are exactly counterbalanced by an equal set of losses.

Economic growth for everyone cannot solve the problem because the demands are not for more but for parity. Yet ours is not a society that believes in absolute equality. Where should parity demands be met and where should parity demands be rejected? What principles should underly this acceptance or rejection? Unless we can learn to answer such questions and implement our answers, our society is going to both stagnate and be split along group lines. There is no way to avoid the problem. Benign neglect will not solve it.

Chapter 8

Solving the Economic Problems of the 1980s

WHEN VIEWED TOGETHER, the problems of the 1980s share both a common set of causes and a common set of cures. Energy, growth, and inflation are interrelated on many fronts. Without growing energy supplies, economic growth is difficult, and rapidly rising energy prices provide a powerful inflationary force. Inflation leads to public policies that produce idle capacity and severely retard growth.

To adjust to a rapidly changing pattern of energy supplies, the energy industry needs to be deregulated. But eliminating regulations, protection, and subsidies is also one of the essential ingredients in any successful program for stimulating economic growth. Because of its value elsewhere in the economy and because it involves the fewest net costs, the elimination of regulations, protection, and subsidies becomes the preferred route to controlling inflation. Upward price shocks are deliberately counterbalanced with planned downward price shocks.

Solving our energy and growth problems demand that government gets more heavily involved in the economy's major investment decisions. Massive investments in alternative energy sources will not occur without government involvement, and investment funds need to be more rapidly channeled from our sunset to sunrise industries. To compete we need the national equivalent of a corporate investment committee. Major investment decisions have become too important to be left to the private market alone, but a way must be found to incorporate private corporate planning into this process in a nonadversary way. Japan Inc. needs to be met with U.S.A. Inc.

A united front cannot be created, however, by simply trying to bulldoze energy, growth, and antiinflation policies down the throats of all those who would be hurt. The losers in this process may not be a majority of the population, but they are certainly large enough to prevent any such policy from being adopted.

A high-quality environment is important, and even if it were not, the time has come to admit that many people think that it is important. Unless we are to be permanently bogged down in fighting about the environment, goods and services simply have to be produced in ways that do not result in environmental deterioration. This means more expensive goods and services. Utility executives may not like stack scrubbers, but it is more important to get coal-fired power plants built than to argue about stack scrubbers. Resistance to those who demand reasonable environmental controls is silly since with a rising standard of living (and it is rising) more and more people are going to move into the economic classes that want a clean environment. Environmentalism is the wave of the future. As such, it makes much more sense for those who are interested in economic growth to reach an accomodation with it than to try to resist it.

If protection, regulations, and subsidies were eliminated, large numbers of individuals would suffer economic losses. If such a policy is ever to be adopted, we have to develop techniques for paying compensation to the individuals who are going to be hurt. Support for failing firms should be minimized, but support for individuals to help them move from sunset to sunrise industries should be generous. It should be generous for the simple reason that if it is not, we will not be able to adopt the policies that the country needs.

Economic growth also means that we must fully utilize the skills and talents of the economic minorities that are now kept out of the mainstream of economic activity. While our economy has survived for a very long time with large income gaps between blacks, Hispanics, American Indians, and women on one side, and white males on the other side, the world has changed and it is difficult to imagine that it can survive as well in the future. But even if it could, old levels of performance are no longer good enough. To reach the levels of productivity reached by others, we have to eliminate this divisive issue. It is not an issue that is going to fade away.

Similarly, we have to stand ready to prevent income gaps between the rich and the poor from increasing. The last twenty years have been marked by success on this dimension, but it will be more difficult to be successful in the next twenty years since we are entering a period of rapidly rising inequalities. Active government involvement in promoting economic growth will also make some Americans richer. We have to ensure that the bottom 60 percent of the population does not fall behind, for if we don't, we won't be able to adopt the growth policies that we need. This means that transfer payments will have to continue to grow for the poor and the elderly, and that our income tax system is going to have to be reformulated to keep the after-tax incomes of the second and third quintiles of our households rising in pace with the rest of the economy.

When one reviews what must be done—massive public investments, budget surpluses to generate more savings, large compensation systems, increases in income transfer payments, and tax cuts for the lower middle class—it is clear that one of the basic ingredients of future progress is a tax system that can raise substantial amounts of revenue fairly. If energy is to be deregulated and the massive income redistributions that are implicit in this policy attenuated, substantial amounts of revenue will be necessary. Some of this may come from taxes on energy—a large excise tax on gasoline consumption—but some of it will have to come from general revenue. If good compensatory systems are to be devised for those who make economic sacrifices in the interests of society, large amounts of general revenue will be necessary. If we are to increase income transfer payments and cut taxes on low-income families that are being squeezed by energy prices and growing inequalities in market earnings, fair taxes will have to be collected from the rest of the population.

At the moment, our tax system is so unfair that it simply isn't capable of doing what is demanded of it. We have to have a tax system that will be perceived as fair. Such a tax system would make it possible to raise the revenue that needs to be raised, but it would also make it possible to adopt the expenditure programs that need to be adopted. More public money for process R&D will make some individuals rich. There is nothing wrong with this if those rich individuals at the same time pay their fair share of taxes.

The need to construct a fair tax system simply emphasizes our primary need. Our society has reached a point where it must start to make explicit equity decisions if it is to advance. The implicit, undefended, unanalyzed equity decisions that have been built into our tax, expenditures, and regulatory policies of the past simply won't carry us into the future. To implement public policies in the future we are going to have to be able to decide when losers should suffer income losses and when losers should be compensated. We have to be able to decide when society should take actions to raise the income of some group and when it should not take such actions. If we cannot learn to make, impose, and defend equity decisions, we are not going to solve any of our economic problems.

The Issue Cannot Be Avoided

Decisions about economic equity are the fundamental starting point for any market economy. Individual preferences determine market demands for goods and services, but these individual preferences are weighted by incomes before being communicated to the market. An individual with no income or wealth may have needs and desires, but he has no economic resources. To make his or her personal preferences felt, he must have these resources. If income and wealth are distributed in accordance with equity (whatever that may be), individual preferences are properly weighted, and the market can efficiently adjust to an equitable set of demands. If income and wealth are not distributed in accordance with equity, individual preferences are not properly weighted. The market efficiently adjusts, but to an inequitable set of demands. It is as if we had an efficient street sweeper who was sweeping the wrong street. To have no government program for redistributing income is simply to certify that the existing market distribution of resources is equitable. One way or the other, we are forced to reveal our collective preferences about what constitutes a just distribution of economic resources.[1]

Both politically and intellectually, our history is one of pretending that we can avoid making explicit decisions about the fair distribution of economic resources. Intellectually, we talk about equal opportunity. Presumably this phrase means that each individual should have an equal chance for economic success, but this still leaves two fundamental questions to be resolved. First, what economic game is to be played—capitalism, socialism—some mixture? Second, whatever game is played, what is to be the distribution of economic prizes to those that win or lose?

Choosing the type of game that you wish to play says little about the optimum structure of prizes, since most economies can be adjusted to produce a wide range of different prizes. Market economies, for example, can exist with or without slavery, with or without public ownership, and with or without economic discrimination. What constitutes a "fair" economic game? Do we let consumers' preferences determine the economic merit of an opera company or do we create, through education, a public demand for operatic performances? Is a fair game a game where each person has an equal chance to win? If chances of winning are to be equalized, do we handicap those born with advantages or compensate those born with disadvantages? What constitutes an equal start? Should every individual be subject to the same initial budget constraint? Consider inheritances. Is there any difference between the individual who inherits one million dollars and the individual whose athletic talents will earn him the same lifetime income?

As these questions indicate, the rules of the natural lottery are not intuitively obvious. The rules can only be specified when one knows the desired distribution of prizes to be generated. Lotteries or market economic games can be formulated to yield any distribution of prizes. The market may be a "fair process" to which most Americans are willing to submit, but it is necessary to stipulate some other principles to determine the equitable distribution of economic prizes within this game.

The problem is identical to that of designing the rules of a football game. To design a fair football game, several decisions need to be made. First, what is the initial starting score? Is it zero-zero or something else? Second, how does one advance the ball and score? Third, how often does the game start over? The answer to none of these questions is axiomatic in either the sporting world or the economic world. At Oxford there is, for example, a rowing race that started only once. Every year boats begin where they left off the year before. The race is never over. Would we define the equivalent game as "equal opportunity" in the economic world? History decides the unequal starting point of each individual economic runner and each economic runner is now allowed to hand in his or her baton to whomever he wishes and at whatever point he wishes. The race never starts over. Once a duke always a duke.

But leaving aside the starting score and the problem of how often do you start over, how would you decide whether the rules of advancing the ball are fair or unfair? Presumably, it has something to do with a determination that players of equal ability have an equal probability of scoring, if not winning. How do you determine this in an economic game as complicated as that of the real economy? If women, for example, who work full-time, full-year earn less than 60 percent of what males earn, and that has been true for the entire forty years that we have kept track of such statistics, does that prove that the rules of advancing the ball are unfair? It is either unfair or you have to be willing to defend the position that women are inferior to men.

As a result, it is not possible to retreat to the position that we should specify the rules of a fair economic game and then let this game determine the fair distribution of purchasing power—an initial score. This requires an equity decision. Many fair games that produce many different distributions of prizes could be constructed. To pick which fair game we wish to play we must decide which distribution of prizes we want. There is no escape from having to make explicit equity decisions.

While it is not possible to pick a value-free, fair economic game, there is another well-traveled route in our attempts to avoid explicit decisions about equity. Surprisingly, this is a route that has been used by both conservatives and Marxists. It is the doctrine of superabundance and satiated wants. In Marx's utopia, there was no need to specify an equitable distribution of economic resources since everyone had everything they could want. Workers had no demands for additional goods and services. If everyone has everything he wants, there is nothing to fight about. The problems of equity, the nation state, and personal budget constraints all withered away.

Conservatives often subscribe to this vision. They simply have a different route for getting from here to there. Instead of proceeding from capitalism, to socialism, to communism, and then to utopia, they focus on economic growth and the process of getting to satiated superabundance. Today's inequalities are justified in terms of their contribution to economic growth and the achievement of economic justice tomorrow. If we just grow fast enough, there is more for everyone and equity problems will disappear.

Unfortunately our demonstrated ability to generate new wants has eliminated the possibility—for both Marxists and conservatives—of ever being able to satiate everyone's wants. Since the problem of unsatiated wants is always with us, the problem of specifying economic equity is always with us.

This had led to a retreat from the doctrine of satiated wants to the doctrine of satiated needs. The goal here is to satiate physiological needs as opposed to the wants that are artificially generated by society. What is the minimum amount of income a person (or family) would need to have a perfectly balanced diet and as long a life expectancy as is medically possible? This is the basic question. But problems arise, since the answer to this question yields a very low poverty line. Consider the cheapest medically balanced diet. By combining soybeans, lard, orange juice, and beef liver (edible, cheap, nutritious, but hardly enjoyable foods) a medically balanced diet can be created that costs less than 80 dollars per person per year (according to 1959 prices).[2] It would be a better diet, medically speaking, than most of us now eat. But are we ready to compel people to eat it? Similarly, how much housing space per person is necessary in order to live to a ripe, old age. The answer—very little. Are we then prepared to ignore the housing wants of poor people?

And what does society do about poor families that are ignorant, inefficient, or stubborn? Does a family have an unmet need if it does not know the cheapest way to have a medically balanced diet? Does a family have an unmet need if it does not want the diet that it knows it should have and can afford? Does a family have an unmet need if it simply refuses to eat an unappetizing or unusual diet?

Since the United States has very few people in poverty, when poverty is based on such a definition of physiological "needs," poverty lines were specified in terms of need—but need itself was defined in a relative manner (that is, in terms of "wants"). Given that a family is going to want to eat as other American families, and given that it is going to manage its resources in the same inefficient manner, how much income does it need to get a medically balanced diet (in spite of itself, if you will)? Given that it is going to want to consume something like the same amount of heat per person, how much heat does it need? But the minute that "needs" are defined in terms of "wants," the concept of need loses its concreteness. Wants become necessities whenever most of the people in society believe that they are in fact necessities. Anything to which we have grown accustomed and that is generally available becomes a necessity. Needs, thus defined, grow right along with average incomes. Like satiated wants, satiated needs will not occur.

This phenomenon can be seen in Gallup polls that have repeatedly asked, "What is the smallest amount of money a family of four needs to get along in this community?" The responses are a rather consistent fraction of the average income of the time at which the question was asked—but the sum grows in absolute terms. The answers to this question indicate that families estimate their own minimal needs to be a little more than half of the average family's consumption of that day. Similarly, when asked to categorize people as "poor, getting along, comfortable, prosperous, or rich," the public rather consistently does so relative to average incomes.[3]

What sociologists call *relative deprivation* is a very real feeling in a liberal democracy. Studies in this area indicate that individuals have a very strong feeling that economic benefits should be *proportional* to costs (that is, efforts, hardships, talents, and the like), but that equals should be treated equally. Since there are various types of such "costs" in any situation, and different rewards (income, esteem, status, power), the problem immediately arises as to how equals are defined and how proportionality is to be determined. This has led to the difficult problem of *reference group* determination. To what group of people do you compare yourself to determine whether you are being treated relatively equally and proportionally?

Reference groups seem to be both stable and restricted by the fact that people look at groups that are economically close to themselves. This explains why inequalities in the distribution of economic rewards that are much larger than inequalities in the distribution of personal characteristics seem to cause little dissatisfaction, and why people tend to ask for rather modest amounts if they are asked how much additional income they would like to be making. The happiest people seem to be those who do relatively well within their own reference group rather than those who do relatively well across the entire population. It also explains why studies find immense anger at the welfare system among working people. Those on welfare are clearly a group where benefits are not proportional to costs. They do not need to incur any costs (make any effort) to receive benefits.

Apart from obvious cases such as welfare, where benefits and costs are out of proportion, our conception of what constitutes proportionality and relative equality tends to be heavily determined by history and culture. Distributions of the past are fair until proven unfair. Great social shocks, such as wars and economic depressions, seem necessary to change specifications of relative deprivation.

This is evident in American history. The only recent periods of rising market income equality in the United States occurred during the Great Depression and World War II. From 1929 to 1941 the share of total income going to the bottom 40 percent of all families rose from 12.5 percent to 13.6 percent, while the share of income going to the top 5 percent fell from 30.0 percent to 24.0 percent, and the share of income going to the top 20 percent fell from 54.4 percent to 48.8 percent. From 1941 to 1947 the share going to the bottom 40 percent rose further to 16.8 percent, while the share going to the top 5 percent fell to 20.9 percent, and the share going to the top 20 percent fell to 43.0 percent.[4]

In the Great Depression, an economic collapse was the mechanism for change. Large incomes simply had further to fall than small incomes. In World War II there was a consensus that the economic burdens of the war should be shared relatively equally ("equal sacrifice"); consequently the federal government used its economic controls over wages to achieve more relative equality. Wage policies during World War II were a manifestation of a change in the sociology of what constituted "fair" wage differentials, or relative deprivation. As a consequence of the widespread consensus that wage differentials should be reduced, it was possible to reduce wage differentials deliberately. After they had become embedded in the labor market for a number of years, these new differentials became the new standard of relative deprivation and were regarded as the "just" wage differentials, even after the egalitarian pressures of World War II had disappeared. The important thing to note, however, is that the new standards were not imposed by government on a reluctant population but were imposed on the market by popular beliefs as to what constituted equity in wartime. No one knows how to engineer such changes in less extreme situations.

Equity Goals

Since equity decisions cannot be deduced from purely factual or logical statements, I am under no illusions that the suggestions that follow will meet with universal approval. At the same time, they illustrate a process that everyone must go through in reaching their own equity decisions.

Let me start by suggesting a possible specification of economic equity. In the United States there is a strong allegiance to the principle that people should fairly compete for a distribution of market prizes. At the same time, there is the recognition that the market has not given everyone an equal chance to win. The group that comes closest to our ideal vision of the natural lottery is composed of fully employed white males. They do not suffer from the handicaps of discrimination, lack of skills, or unemployment. If we look at their earnings rather than their income, inherited wealth plays a relatively small role in their current position.

Let me suggest that our general equity goal should be to establish a distribution of earnings for everyone that is no more unequal than that which *now* exists for fully employed white males [5] (see table 8–1). Since this distribution of earnings is the current incentive structure for white males, there are no problems with work incentives. With more than half of the labor force (measured in hours of work) now participating in this natural lottery, it is hardly a distribution of economic resources that anyone could consider un-American.

TABLE 8–1

Distribution of Earnings in 1977

	Full-Time, Full-Year	
Quintiles	White Males (%)	All Other Workers (%)
1	7.7	1.8
2	13.9	7.2
3	18.2	15.8
4	23.5	27.0
5	36.7	48.2
Mean Earnings	$16,568	$5,843

SOURCE: U.S. Bureau of the Census, *Current Population Reports, Consumer Income 1977,* Series P–60, no. 118 (March 1979), p. 228.

Unless one believes that the culture in which women, minorities, and underemployed white males exist is different from the culture in which employed white males exist, there is every reason to believe that a reward structure that is capable of keeping white males on their economic toes is also capable of keeping other Americans on their economic toes. Inequalities greater than those that now occur in the earnings of fully employed white males are not necessary to keep the economy functioning. They are, in fact, counterproductive.

As is clear from table 8–1, the mean earnings of fully employed white males are about three times as high as those for the rest of the labor force. At the same time, the distribution of income among fully employed white males is much more equal than that for the rest of the population. The earnings of the top quintile of fully employed white males is five times as large as that of the bottom quintile, but for the rest of the population the same gap is twenty-seven to one—more than five times as large.[6] The normal arguments against more equality are couched in terms of the need to sustain work effort, but fully employed white males keep working with a five to one gap. Why should the rest of the population require a twenty-seven to one gap?

While we could probably argue about whether a five to one gap is equal enough to constitute economic justice, it is so far from where we are that we could use it as our interim equity goal for a long time before we had to worry about what else, if anything, should be done. Since many individuals only want to work part time, achieving this goal would not lead to an overall distribution of earnings of five to one, but if everyone had an opportunity to participate in the "natural" economic lottery enjoyed by white males, the distribution of earnings would be much more equal than that which now exists.

How can we go about organizing a society where everyone gets to play the same economic game as that of fully employed white males? The standard economic solution has been to attempt to equalize the distribution of human capital. In a simple supply-demand auction market for laboring skills, this would be the correct solution. If you pump a more equal distribution of skills and productivities into the economy, a more equal distribution of earnings must flow out of the economy.

But there are a number of problems with this solution. The first is that empirically it isn't working. If human capital is measured in terms of education, as it usually is, by any measure we have substantially equalized the distribution of human capital since World War II. Yet as we have seen, the distribution of earnings has become more unequal. This is true for both the entire labor force and for white males.

If you think back to the four replacement hypotheses that allowed us to understand rigid wages, the reasons for the failure became clear. Education is an important background characteristic that affects our costs of acquiring job skills, but by itself it is seldom a productive skill. Working skills and associated earnings are learned on the job. The allocation of jobs determines the allocation of skills and hence the allocation of earnings.

Economic minorities will never catch up with white males unless they have an equal opportunity at the job opportunities open to white males. Reshuffling the current structure of job opportunities might bring equality between different groups, but it would not meet our equity goal. The basic problem is to change the structure of the economy so that the entire economy generates the kinds of jobs that are now open to white males and ensures that there are enough of these job opportunities to go around.

Controlling inflation without idle capacity is essential since we now start from a position where there simply aren't enough jobs, good or bad, to go around. The problem is not just peculiar to this period of stagflation. Lack of jobs has been endemic in peacetime during the past fifty years of American history. Review the evidence: a depression from 1929 to 1940, a war from 1941 to 1945, a recession in 1949, a war from 1950 to 1953, recessions in 1954, 1957–58, and 1960–61, a war from 1965 to 1973, a recession in 1969–70, a severe recession in 1974–75, and another recession probable in 1980. This is hardly an enviable economic performance.

While monetary and fiscal policies could be used to stimulate the economy to the degree that it would provide good jobs for everyone able and willing to work, macroeconomic policies will not be used for this purpose. The reasons are many—fears of more inflation, time lags in decision processes, incompetence—but whatever the reasons, we need to face the fact that our economy and our institutions will not provide jobs for everyone who wants to work. They have never done so, and as currently structured, they never will. When it comes to unemployment, we are consistently the industrial economy with the worst record.

As a result, the principal way to narrow income gaps between groups is to restructure the economy so that it will, in fact, provide jobs for everyone. Since we regard the United States as a *work ethic* society, this restructuring should be a moral duty as well as an economic goal. We consistently preach that work is the only "ethical" way to receive income. We cast aspersions on the "welfare" society. Therefore we have a moral responsibility to guarantee full employment. Not to do so is like locking the church doors and then saying that people are not virtuous if they do not go to church.

Since private enterprise is incapable of guaranteeing jobs for everyone who wants to work, then government, and in particular the federal government, must institute the necessary programs. No one should attempt to deny that a real, open-ended, guaranteed job program would constitute a major restructuring of our economy. Patterns of labor market behavior and the outputs of our economy would be fundamentally altered.

It should be pointed out, however, that real economic competition would almost certainly increase. If the guaranteed jobs are to be real jobs, then any guaranteed job program must produce some economic outputs. These outputs might consist of street cleaning in competition with public sanitation departments, or the rebuilding of railway roadbeds in competition with private industry. The problem is not finding worthwhile things to do. Anyone with even a little imagination can think of many things that could be done to make this society a better one. If the option is between idleness and work, the choice is simple. As long as any useful output is produced, a work project takes precedence over involuntary unemployment.

A guaranteed job program must have several characteristics in order to achieve the objectives for which it is intended. First, it cannot be a program of employment at minimum wage rates. The objective is to open to everyone a structure of economic work opportunities equivalent to those open to fully employed white males. Thus, the program would have to structure earnings and promotion opportunities in the same way as they are structured for fully employed white males. There would be some low-wage jobs and some high-wage jobs, but most jobs would be in the middle. Some or all of the workers might be unionized. Second, the program must be open-ended, providing jobs to everyone who is able and willing to work regardless of age, race, sex, or education. Abilities and talents will play a role within the distribution of job opportunities, but no one who desires full- or part-time work will be denied it. Third, the program should not be viewed as a temporary antirecessionary measure. The lack of employment opportunities is not a temporary, short-run aspect of the U.S. economy. It is permanent and endemic. Even if this were not true, the program would still need to be permanent, since no industry could be expected to go in and out of business over the course of the business cycle and still operate efficiently. There is no reason, however, why there could not be temporary, short-run jobs for people who, for a limited period, are unable to find work in the conventional private or public sectors of the economy.

If the guaranteed job program were structured to provide the kinds of job opportunities now open to white males, the private economy would have to adjust. In our current economy, we can play two different economic games at the same time, since most people aren't allowed to play the primary game. If the primary game were open to everyone, everyone would abandon the secondary game or it would have to transform itself to provide the same working conditions and opportunities. In the face of competition and threats to its own survival, I have no doubt that it would transform itself.

What would such a program cost? Payments for labor, materials, and capital might be high, but as with all economic projects, the costs would depend upon the difference between the value of output produced and the payments made to factors of production. If care is shown in project selection, there is no reason why the projects could not generate substantial net benefits. If you are employing idle economic resources (workers without jobs), the real economic costs (opportunity costs) would be substantially less than the monetary costs.

How many people would need to be employed in a guaranteed job program? If six million people were unemployed, the answer is obviously many millions. If a guaranteed job program were actually instituted, however, the performance of macroeconomic policy makers would improve. If policy makers did not implement fiscal and monetary policies to ensure that private industry would want to hire most of the U.S. labor force, then they (governments) would be forced to hire directly all the people who were left over. I suspect that interest in maximizing private employment opportunities would suddenly arise. The number who needed employment would be large, but much smaller than those currently unemployed.

One of the subsidiary benefits of a guaranteed job program is that it would eliminate the endless sterile debates about what fraction of the unemployed are lazy and unwilling to work. Instead of arguing about it, let's put it to the test and see once and for all how many people really want to work and how many people are, in fact, lazy.

Historically, one of the interesting things about our economy and political structure is that we find it much easier to set up welfare programs to give people money than we do to set up work programs to give people jobs. Transfer payments stood at $224 billion in 1978 while only $10 billion was spent in subsidized jobs.[7] Yet public rhetoric would lead one to believe the opposite. If an equitable distribution of economic resources is ever to be achieved, it will require the provision of jobs for everyone who wants to work.

Politically, we are reluctant to give jobs, because to do so would require a major restructuring of the economy. A new source of competition would arise for both public agencies and private firms. To the extent that we were unable or unwilling to hold the private economy at the full-employment level, we would have a socialized economy.

The time has come, however, to admit that the pursuit of equity and equal economic opportunity demands a fundamental restructuring of the economy. Everyone who wants to work should have a chance to work. But there is no way to achieve that situation by tinkering marginally with current economic policies. The only solution is to create a socialized sector of the economy designed to give work opportunities to everyone who wants them but cannot find them elsewhere.

I am not naïve enough to think that such a plan is about to be adopted, but the basic problem is already of long standing. Can you really imagine continuing for another thirty years with black unemployment twice that of whites? Full-time, full-year female workers have earned less than 60 percent of men ever since record keeping began more than forty years ago. Can it continue for another forty years? Perhaps, but I doubt it. Should it continue?

I have no problem answering in the negative. Will it continue if we don't do something to change the structure of the economy fundamentally? I have no problem answering in the positive. Will we fail as a society to address this fundamental problem and let it drag us down with it? Perhaps. In the records of history, we certainly would not be the first society that failed to come to grips with its fundamental internal problems.

If we were to achieve a distribution of earned income opportunities for the entire population, such as that now enjoyed by fully employed white males, much of the welfare problem would disappear. But there would still be a problem among the elderly and those families without earners for medical or other reasons. In the case of the average elderly person, the problem is maintaining the income parity that the social security system is now providing. For the poor, elderly, or otherwise, the aim should be to establish an income transfer payment system which provides a standard of per capita living that is approximately half that enjoyed by the rest of the population. This is what the public opinion polls seem to indicate that we want. Interestingly, it is also almost exactly what the official poverty line specified when it was first established in 1963 and what many northern and western states were providing in welfare in the early 1970s.

Our tax reform goals should focus on two principles. First, wherever possible, reforms should be taken to reduce the dispersion in tax rates. Proposals that increase the dispersion in tax rates for those in the same income class should be avoided. Whatever the right degree of vertical equity, horizontal equity (the equal taxation of equals) is an important principle that our current tax system cruelly violates.

Second, the appropriate degree of vertical equity depends upon how closely we come to achieving an equitable distribution of market earnings. If we reached a distribution of market earnings in accordance with that suggested above, a proportional tax system would be appropriate. To the extent that we have not achieved an equitable distribution of market earnings, the tax system should be structured to move whatever distribution of market earnings does exist toward an after-tax distribution of income that approaches our equitable distribution of market incomes.

In a transition period, progressive taxes should be levied to move us toward our general equity goals, but taxes are an inferior secondary approach to the problem. They are to be used only until more fundamental changes can be put in place. The goal is not a system of taxes and transfers that leads to an equitable distribution of economic resources, but a system of market earnings that is equitable.

Once we agree on a specification of economic equity, we are in a position to deal with the problem of economic change and compensate those that get hurt when public policies are altered. When should temporary adjustment assistance be paid to economic losers, and when should it not be paid? One of the peculiarities of our mixed economy is that we have poor to nonexistent systems for compensating individuals who legitimately lose when projects are undertaken in the general interest. The only recourse of individuals in this situation is to stop the economic progress that threatens them. If we want a world with more rapid economic change, a good system of transitional aid to individuals that does not lock us into current actions or current institutions would be desirable.

Adequate individual compensation systems are opposed for a number of reasons. Sometimes compensation would have to be paid to those who are already rich compared with the rest of the population. Since the rich are seen as avoiding their fair share of the tax burden, compensation for losses is seen as doubly unfair. The appropriate correction for this problem is not to resist compensation systems but to establish a fair tax system. Project developers (government or private) are used to getting what they want without compensation, and they resent having to pay. To pay compensation is to admit that governments and firms have income distribution responsibilities. Incomes do not go up and down because of impersonal market forces. And since many factors cause incomes to go up and down in a large economy, it is difficult to decide when compensation should or should not be paid. Not all losses can or should be compensated.

All of these objections have merit, but even together they do not constitute an adequate case against compensation. We simply need a better system if we are to have any hope of making the economy more dynamic. If rich people are hurt and are rich in accordance with our specification of economic equity, they deserve to be compensated just as much as the poor. Project managers may resent having to pay compensation, but the current choice is between paying compensation or never getting the project underway. Government does have income distribution responsibilities, and we simply aren't economic fatalists anymore. Any compensation system will, to some extent, be arbitrary and fail to help someone that it should help; however, a second-best system is not perfect but better than no system at all.

Existing compensation systems simply illustrate the problems. Instead of being run as if they were intended to give generous compensation for losses actually suffered, they are run as if the aim is to deprive the citizen of his income or capital. Parsimony rather than generosity is the rule. In urban renewal, compensation is paid for property and moving expenses, but a very narrow interpretation is taken of what constitutes a loss. No compensation is paid for disrupting lives or for the loss of neighborhoods–friends, comfortable habits, and so forth. These losses are undoubtedly difficult to quantify, but they are nonetheless real. Not being willing or able to quantify them precisely, we act as if they were not losses at all. The same approach is followed in the Trade Adjustment Assistance Act. Since the benefits of free trade are general, while the costs are usually localized, it would seem fair to compensate the losers from the general gains. Yet until recently, adjustment assistance has been run as if the aim were not to spend any money or find any cases of valid disruptions and losses. Administratively, the programs are often even less generous than they seem on paper.

To conduct either public or private business, more adequate compensation systems are going to have to be developed in the future. Those who suffer the localized costs that generate universal benefits are going to have to be compensated. But this is also likely to make a change in the mixture of the mixed economy, since government will undoubtedly be called upon to help decide what constitutes compensation and how the necessary revenue should be collected. If we cannot develop better compensation systems, then recommendations to end protection, subsidies, and price controls are useless. Individuals want economic security, and to simply recommend that they give it up is to shout at the wind.

There are two possible avenues. One avenue is that followed by the Japanese, where large firms are deliberately structured on a conglomerate basis and then helped to shift resources from sunset areas to sunrise areas within the same firm. Individuals are protected since they know that they will also be transferred from the sunset areas to the sunrise areas. The other avenue is that followed by Sweden, where an attempt is made to provide an individual safety net. An elaborate social welfare system cushions economic shocks, and large amounts of resources are used in retraining the work force to move from one area to another.

Given that the private economy already seems to be moving toward a conglomerate form of organization, this should be encouraged and developed into a system that promotes productivity and economic security. Instead of prohibiting mergers, firms should be encouraged to engage in different activities. With our pattern of heavy internal financing from retained earnings and depreciation allowances, investment funds are much more apt to flow into high-productivity areas if managers can invest the funds in their own firm. Like the rest of us they want economic security and tend to reinvest in low-productivity areas if that is the only way to protect their economic security. This would improve the allocation of both internal savings and external funds. External lenders are only interested in being repaid, and often low-productivity borrowers, such as the steel industry, are very safe risks because of their large internal savings. If the internal funds are more efficiently allocated, the external funds will automatically be more efficiently allocated. Firms should be encouraged to move into new areas and out of old areas, but only with the understanding that they are expected to take their workers, as well as their managers, with them.

The individual safety-net approach also needs to be used. Transitional aid for retraining, relocating, and getting through a period of unemployment should, if anything, be overly generous. The goal is not to spend the least possible, but to promote a rapid rate of economic change. What we lose in overly generous compensation, we will more than make up in faster economic change.

What needs to be avoided is the institutional, safety-net approach where firms are protected from failure in the name of protecting individuals. Wage and retraining subsidies should be attached to individuals. They may be cashed by firms who employ these individuals, but subsidies should not be given to firms directly. There is a sharp distinction to be made between protecting the failing individual and protecting the failing firm.

Decisions to Be Made

Whatever the process for getting there, and whatever the specifications of economic equity, there are four major decisions that everyone must make.

First, what is the minimum economic floor to which you will let any individual or family sink regardless of the cause of their failure? Unless you are willing to tolerate starving families in the streets, this is a question that must be answered by everyone. I suggest a minimum floor that would provide a standard of living just half as large as that of the average American.

Second, what is to be the distribution of economic rewards for those that participate in the economy? I suggest that structure of rewards that now exists for fully employed white males.

Third, given that tax revenue must be collected to finance government expenditures, how should this burden be distributed? Given a fair distribution of economic rewards in the marketplace, a proportional tax system is desirable, but without large variances among individuals with the same real income. To the extent that the distribution of market earnings has not reached the desired level, a progressive tax system should exist to move the distribution of take-home incomes toward the desired goal.

Fourth, what compensatory payments should be made when public policies cause large income losses? One can be a purist and answer "never," but I argue that we need a generous system of transitional aid to individuals, but not firms.

The Political Process

While this has been a book on economic problems, these problems and their solutions focus attention back on our political process. Does our inability to act reflect fundamental irreconcilable divisions that no political process could overcome, or is there something wrong with our political system? Some of our paralysis is due to irreconcilable differences, but some of it is also because of a political process that cannot make decisions when all decisions result in substantial income losses for someone.

This is not a fault shared by other forms of government to the same extent. Everywhere else in the industrial world, parliamentary forms of government have demonstrated that they can penalize automobile driving, even when everyone drives and loves it. Very high taxes can be levied on gasoline elsewhere, but not here.

Our problems arise because, in a very real sense, we do not have political parties. A political party is a group that can force its elected members to vote for that party's solutions to society's problems. With a majority and minority party, the majority is expected to solve the nation's economic problems. If it can't, it is replaced in the next election, and the minority becomes the majority. Responsibility for success is clear, and failures can be punished.

Instead of having two parties, we have a system where each elected official is his own party and is free to establish his own party platform. Parties are merely vague electoral alliances. But this means a splintering of power that makes it impossible to hold anyone responsible for failure. No elected official can be expected to solve the problems by himself. Failure can always be blamed on someone else. There is no majority that must solve problems or be held accountable. In comparison, the diffusion of responsibility that we so often castigate in proportional representation seems mild. We have the ultimate in proportional representation, where every elected official is a one-man political party.

When no one can be held responsible for failure, it becomes possible for everyone who contributed to the failure to be reelected time after time. Each individual member of Congress reports to his constituents that his solutions to the problems were killed by someone else, but he is fighting hard. He or she can also report that they also successfully fought to prevent their particular electorate from having to suffer any of the costs that would have occurred if someone else's solutions to the problems had been adopted. Being successful in stopping programs that would hurt their electorate, and giving the appearance of working toward solutions, each congressman can be reelected with the problems unsolved. Since no one has the power to solve the problems, no one can be fired for not solving the problems.

But not having accountable, integrated political parties fails us in an even more fundamental way. Since all economic solutions require decisions about the distribution of income, we should be voting political parties up or down based on how they are going to allocate the economic losses necessary to solve our problems. Not having political parties with a common position on this issue, there is no way that voters can come to a majority or minority position on who should bear the inevitable losses. Each individual congressman is free to argue that all of the losses should be allocated to someone else's congressional district, and this is exactly what his voters want to hear.

Presidential candidates cannot shift the losses to someone else's electoral district quite so easily; therefore they retreat to the position that they can solve the problems without hurting anyone. We are told about the large economic gains that each of us will make if they were elected, but losses either don't exist or are quietly ignored.

To pretend that there are no losses, however, is to guarantee that once elected, a president will not be able to impose the necessary losses. He has been elected on the basis of no losses for anyone, and he has no electoral mandate to impose the losses. In contrast, a British conservative government was elected on the platform of tax cuts for the rich and tax increases for the lower middle class. Having been elected on this redistribution platform, the laws implementing it could be quickly passed. In our system, proposals that yield economic losses come as a surprise, are treated as a betrayal, and result in fierce political resistance that makes it impossible to impose the programs.

There is no easy path for getting from here to there, but somehow we have to establish a political system where someone can be held responsible for failure. This can only be done in a system where there are disciplined majority and minority parties. Every politician with his or her own platform is the American way, but it is not a way that is going to be able to solve America's economic problems.

As we head into the 1980s, it is well to remember that there is really only one important question in political economy. If elected, whose income do you and your party plan to cut in the process of solving the economic problems facing us? Our economy and the solutions to its problems have a substantial zero-sum element. Our economic life would be easier if this were not true, but we are going to have to learn to play a zero-sum economic game. If we cannot learn, or prefer to pretend that the zero-sum problem does not exist, we are simply going to fail.

Notes

Chapter 1

1. International Monetary Fund, *International Financial Statistics* 32, no. 4 (April 1979): 122, 156, 214, 352, 356, 390.
2. Ibid., p. 228.
3. Irving Kravis, Alan Heston, and Robert Summers, "Real GDP for More than 100 Countries," *Economic Journal,* June 1978, p. 215.
4. International Monetary Fund, *International Financial Statistics* 32, no. 4 (April 1979): 43, 354.
5. Ibid., pp. 154, 214, 390.
6. United Nations, *Yearbook of National Account Statistics, 1977,* vol. 1 (New York: United Nations, 1978), p. 348.
7. Malcolm Sawyer and Frank Wasserman, "Income Distribution in OECD Countries," *OECD Economic Outlook,* July 1976, p. 14.
8. U.S. Department of Commerce, Bureau of Economic Analysis, *The National Income and Product Accounts of the United States, 1929–1974,* p. 312.
9. U.S. Department of Commerce, *Survey of Current Business* 59, no. 7 (July 1979): 43.
10. Richard Easterlin, "Does Money Buy Happiness?" *The Public Interest,* no. 30 (Winter 1973), p. 3.
11. See Chapter 7.
12. Edward E. Lawler II, *Pay and Organizational Effectiveness: A Psychological View* (New York: McGraw-Hill, 1971), p. 37.

Chapter 2

1. U.S. Department of Commerce, *Survey of Current Business* 59, no. 7 (July 1979): 35.
2. U.S. Department of Commerce, *Statistical Abstract of the United States* (Washington, D.C.: U.S. Government Printing Office, 1978), p. 488.
3. Calculated on the assumption that all energy prices would have risen proportionally to that of oil on a BTU basis.
4. See Table 2–1.
5. U.S. Federal Reserve Board, "Survey of Financial Characteristics of Consumers," *Federal Reserve Bulletin* March 1964, p. 285.

6. Robert S. Pindyck, *The Structure of World Energy Demands* (Cambridge, Mass.: MIT Press, 1979), p. 43.

7. For a discussion of energy and the environment see Chapter 5.

Chapter 3

1. See Table 2–1.

2. U.S. Department of Commerce, Bureau of Economic Analysis, *The National Income and Product Accounts of the United States, 1929–1974* (Washington, D.C.: U.S. Government Printing Office, 1975), p. 264.

3. Ibid.

4. U.S. Department of Commerce, *Statistical Abstract of the United States* (Washington, D.C.: U.S. Government Printing Office, 1978), p. 713.

5. U.S. Department of Commerce, *Survey of Current Business* 59, no. 7 (July 1979): 26.

6. Ibid.

7. Council of Economic Advisers, *Economic Indicators,* Sept. 1979, p. 6.

8. Ibid., p. 2.

9. See Table 3–2.

10. U.S. Department of Commerce, *Survey of Current Business* 59, no. 7 (July 1979): 39.

11. U.S. Bureau of the Census, *Current Population Reports, Consumer Income 1977,* Series P–60, no. 119 (March 1979), p. 5.

12. U.S. Bureau of the Census, *Current Population Reports, Consumer Income 1977,* Series P–60, no. 117 (Dec. 1979), p. 34.

13. Franco Modigliani and Richard A. Cohn, "Inflation, Rational Valuation, and the Market," *Financial Analysis Journal* 10 (March–April 1979): 3.

14. U.S. Department of Labor, *Formal Occupational Training of Adult Workers,* Manpower Automation Research Monograph, no. 2 (Washington, D.C.: U.S. Government Printing Office, 1964), p. 201.

15. Richard Easterlin, "Does Money Buy Happiness?" *The Public Interest,* no. 30 (Winter 1973), p. 3.

16. International Monetary Fund, *International Financial Statistics* 32, no. 12 (Dec. 1979): 165, 369.

17. "West's Inflation Rate Found Accelerating," *New York Times,* June 13, 1979, p. D-5.

18. See Table 3–3.

19. Based on relative unemployment rates in Table 3–3.

20. U.S. Department of Commerce, *Survey of Current Business* 59, no. 7 (July 1979): 39.

Chapter 4

1. Joint Economic Committee, *Manufacturing Productivity Growth, 1960–77* 5, no. 7, p. 1.

2. See Chapter 3.

3. Council of Economic Advisers, *Economic Report of the President,* Jan. 1979, p. 226
4. U.S. Department of Commerce, *Statistical Abstract of the United States* (Washington, D.C.: U.S. Government Printing Office, 1978), p. 622.
5. U.S. Department of Commerce, *Survey of Current Business* 59, no. 7 (July 1979): 26.
6. U.S. Department of Commerce, *Survey of Current Business* 59, no. 7 (July 1979): 52, 56.
7. See Chapter 8.
8. U.S. Bureau of the Census, *Current Population Reports, Consumer Income 1977,* Series P–60, no. 117 (Dec. 1978), p. 19.

Chapter 5

1. Council of Economic Advisers, *Economic Report of the President,* Jan. 1979, p. 226.
2. Ibid., pp. 246, 209.
3. Ibid.
4. See Chapter 2.
5. Council of Economic Advisers, *Economic Report of the President,* Jan. 1979, p. 184.
6. Calculated by regressing relative income on the national unemployment rate.
7. U.S. Department of Labor, *Employment and Earnings* 26, no. 1 (Jan. 1979): 181.

Chapter 6

1. Charles B. Burck, "The Pros and Cons of Deregulating the Truckers," *Fortune* 99, no. 12 (July 18, 1979): 140.

Chapter 7

1. U.S. Department of Commerce, *Survey of Current Business* 59, no. 7 (July 1979): 16.
2. See Table 7–1.
3. Council of Economic Advisers, *Economic Report of the President,* Jan. 1979, pp. 268, 269.
4. See Table 7–3.
5. See Table 7–2.
6. U.S. Department of Commerce, *Survey of Current Business* 59, no. 7 (July 1979): 44.
7. See Table 7–4.
8. See Table 7–4.

9. These numbers are calculated from the same Census tapes as the data presented in table 7–4.

10. See Table 7–6.

11. See Table 7–7.

12. Daniel M. Holland, "The Effects of Taxation on Effort" (Paper at the proceedings of the 62nd National Tax Association Oct. 1969), p. 428.

13. Joseph A. Pechman and Benjamin A. Okner, *Who Bears the Tax Burden?* (Washington, D.C.: The Brookings Institution, 1974), p. 46.

14. U.S. Bureau of the Census, *Current Population Reports, Consumer Income* 1977, Series P–60, no. 118 (March 1979), p. 45.

15. See Table 7–8.

16. U.S. Department of Commerce, *Survey of Current Business* 59, no. 7 (July 1979): 39, 40.

17. George Cooper, *A Voluntary Tax? New Perspectives on Sophisticated Estate Tax Avoidance* (Washington, D.C.: The Brookings Institution), 1979, p. 40.

18. Arthur Louis, "The New Rich," *Fortune* 88, no. 3 (Sept. 1973): 170.

19. Two excellent survey articles of the random walk are Eugene F. Fama, "Efficient Capital Markets: A Review of Theory and Empirical Works," *Journal of Finance,* no. 25 (May 1970), pp. 383–417; and Michael C. Jensen, "Capital Markets: Theory and Evidence," *The Bell Journal of Economics and Management Science,* no. 3 (Autumn 1972), pp. 357–398.

20. Ibid.

21. Jacob Mincer, *Schooling, Experience, and Earnings* (New York: National Bureau of Economic Research, 1974), p. 112.

22. U.S. Bureau of the Census, *Current Population Reports, Population by Ethnic Origin 1972,* Series P–20, no. 249, p. 26.

23. U.S. Department of Labor, *Employment and Earnings* 26, no. 1 (Jan. 1979): 156.

24. U.S. Bureau of the Census, *Current Population Reports, Consumer Income 1977,* Series P–60, no. 118 (March 1979), p. 234.

25. Ibid.

26. Ibid.

27. U.S. Department of Labor, *Employment and Earnings* 26, no. 1 (Jan. 1979): 189.

28. U.S. Bureau of the Census, *Current Population Reports, Consumer Income 1977,* Series P–60, no. 118 (March 1979), pp. 218, 222.

29. U.S. Bureau of the Census, *Current Population Reports, Persons of Spanish Origin,* Series P–20, no. 339 (March 1978), p. 27.

30. U.S. Bureau of the Census, *Current Population Reports, Consumer Income 1977,* Series P–60, no. 118 (March 1979), pp. 227, 231.

31. U.S. Department of Labor, *Employment and Earnings* 26, no. 1 (Jan. 1979): 156.

32. U.S. Bureau of the Census, *Current Population Reports, Consumer Income 1977,* Series P–60, no 118 (March 1979), pp. 197, 198.

Chapter 8

1. Economic theory avoids equity decisions by retreating into what is called Pareto efficiency–a fancy term for "more is better than less." If a public program moves the economy from State *A* to state *B,* and in state *B* everyone is better off than, or as well off as, they were in state *A,* then we can say that the public policy is Pareto-efficient and should be adopted. But since there is always

Notes

someone who is worse off after any change, nothing is Pareto-efficient in the real world. As a result, we retreat farther to the weak form of Pareto efficiency. In this weak form, state *B* is Pareto-efficient if the economic gainers in state *B* could compensate the economic losers in state *B* so that everyone is as well off or better off. This, of course, is always possible as long as total resources in state *B* are larger than in state *A*. Therefore any policy that raises the GNP is Pareto-efficient. The problem in the real world is that the compensation from winners to losers actually has to be paid, yet is almost never paid. As a result, we cannot avoid making economic equity decisions in public policies, even though we can eliminate them in economic theory.

2. Victor E. Smith, *Electronic Computation of Human Diets,* M.S.U. Business Studies (E. Lansing: Michigan State University, 1964), p. 20.

3. Lee Rainwater, "Poverty, Living Standards and Family Well-being," Harvard-MIT Joint Center for Urban Studies Working Paper no. 10, p. 45.

4. Herman Miller, *Income Distribution in the United States* (Washington, D.C.: U.S. Bureau of the Census, 1966), p. 21.

5. See Table 8–1.

6. Ibid.

7. U.S. Department of Commerce, *Survey of Current Business* 59, no. 7 (July 1979): 16.

Index

Index

Index

Index

Index